投考公務員AO/EO

增 修 第 五 版

JRE全攻略

出題模式｜必溫重點｜政策分析

李Sir 著

目錄

序言

JRE 唔知溫乜？

JRE（Joint Recruitment Examination，聯合招聘考試）很難考？主觀來說，這個考試非常易過關（中、英文卷各取100/200分或以上）。筆者畢業於香港一級學府，班中大部分學生（包括筆者）應考JRE均是「1 take pass」，倒有不少同學是面試後 reject offer，或在三年試用期內辭去公務員職位。

很多朋友向我表示，他們考了多次JRE均失敗，因為「唔知溫乜」。一般中、小學生都是說「唔知點溫」，但「唔知溫乜」卻嚇了我一跳。不知道讀什麼，便拿本JRE的Textbook讀一篇便可。

怎知很多朋友原來都不知道JRE有Textbook，想當年筆者在一級大學讀書時，是有credit-bearing的課程教授應考公務員（當年課程名稱有點模糊），亦有JRE的Textbook、各大政府職系的面試題目及評分標準等……不知道這些東西是失傳了，亦或者現在年青人只懂用 Google 找資料，上網找不到便說沒有。

本書面世前，我讓幾位朋友（分別是 repeat、tripeat、quadpeat、hexapeat，甚至 legendary-peat JRE）先看一次，他們均表示「從來唔知咁樣考、咁樣溫」，希望此書能幫助各考生備戰。

JRE 溫習模式

在筆者的讀書年代，教授提供了一套架構系統的 JRE 溫習模式，怪不得 AO 成了香港一級學府舊生會。為了讓其他大學學生能公平參與 JRE 考生，筆者特此著書分享 JRE 溫習模式。

1. 英文試卷練 Data-based question

英文試卷考的是學生應用資料能力，讀者可在模擬試卷中看到，考生要在短時間整合大量資料，若無事先預備，根本難以合格，所以本書備有多個答題技巧，助考生理解資料。

2. 英文文法不可錯

一個考生若文法有大錯漏，基本上是被「即時 DQ」無誤。本章的文法章節不是談及基本文法，而是句構變化。要取得可面試的分數，除了試卷內容命中評分準則外，亦要有一定的句構變化。

3. 中文試卷講求通識

每年 JRE 的中文題都嚇親考生，有一年考難民營、有一年考郵局增設匯款功能等，考生若無一定的時事知識，考官「一睇便穿煲」。本書「爆料」 50 個考試官出題庫，可以幫助考生預先對各個議題有初步認識。

一些坊間劣質 JRE 班，只會派發考試臨近的新聞摘要予考生參詳，此做法極不可取，考生應讀的是考官和政府關注的50個議題。本書已逐一列舉。

最後祝各考生「1 take pass」。

CHAPTER 1

英文試卷剖析

背景介紹

很多考生都忽視了第一段的背景介紹。這是説明個案分析的地區並非香港，但其社會特徵（demographic character）則與香港情況極為相似。考生在回應考題時，切記要加插相關為資料。

例題：

Manmantrlia is a resilient, vibrant and dynamic economy in Asia. The ability to sustain the prosperity of the economy is dependent on the attractiveness of a low and competitive tax system. The corporate and personal income tax rates are among the lowest in the world. However, Manmantrlia currently rely upon only a very limited taxes to generate revenue to support the expenditure.

題目：You are the officer responsible for tax reform. Please draft a discussion paper for the Prime Minister's setting out the pros and cons of raising the tax rate of corporate and personal income.

參考答案（壞處）：Once the reform is implemented, Manmantrlia's tax arrangements may not compete with other asset management centres. For example, Singapore, one of the major financial centre in Asia, has implemented a number of tax incentives to attract fund management companies.

考試技巧：很多考生忽略 Asia 這字，導致失分，亦有考生誤以為題目所說的是香港，於是加插香港的相關資料，例如薪俸稅階梯等，這亦是考生常犯的錯誤。考生必須提醒自己，不要胡亂加插香港的情況作答，否則會違反了 Data-based question 的原意，難以合格。

數據描述(趨勢)

數據題是多個國際考試都會採納的題型，例如雅思（IELTS），考生經常忽略了兩項重要元素，便是計數機和副詞。第一，考生必須攜帶計數機，以計算數據的變化；第二，考生需用大量副詞去形容數據的轉變，不能只用 increase/ decrease/ raise/ drop 這麼簡單，要寫的是 increase sharply/ decrease significantly/ raise slightly/ drop lightly 等。

例題：Figures of unemployment rate in Nicnicland 2016-2020

Year	2016	2017	2018	2019	2020
Unemployment rate (%)	15%	12%	10%	10%	5%

題目：You are the officer responsible for alleviating the unemployment problem in Nicnicland. Please draft a discussion paper for the Prime Minister's labour market committee, setting out the pros and cons of introduction of training programme in the labour department.

參考答案（講解情況）：The unemployment rate in Nicnicland declined between 2016 and 2020, a period of 5 years. In 2016, the rate was 15%. It dropped slightly to by 3% to 12.5% and by 2% to 10.5% in 2017 and 2018 respectively. The rate remained unchanged between 2018 and 2019 but it decreased by 2 times to 5% in 2020.

考試技巧：在討論培訓計劃的優點或缺點前，考生應試描述現今的勞工市場情況。在引述數據方面，不少考生不知道要帶計數機，但有的考生雖然帶了計數機，卻又忘了把原本的數據抄上去。

其實應付這些數據題記住兩個步驟便可：先抄數據，然後計算。計算方面，著量要用乘號（×）、除號（÷），甚至百分比（%），以確保可在這部分取滿分。

數據描述(特徵)

數據題除了描述特徵描寫外，考生有時亦需留意數據的特徵，這方面需要較高的英語水平詞匯。以下列出一些常用字詞：

- Marginal（邊際）
- Consistent（穩定）
- Overall（總括而言）
- Polarization（兩極化）
- Scattering（分散）

例題：The annual median income of top 10% and bottom 10% families in Boyboyland

Year	2018	2019	2020
Top 10% family annual medium income (BB$)	50,000	80,000	100,000
Bottom 10% family annual medium income (BB$)	10,000	8,000	6,000

題目：You are the officer responsible for alleviating the income disparity between rich and poor in Boyboyland. Please draft a discussion paper for the Prime Minister's economic bureau, setting out the pros and cons of the increase in the progressivity of the salaries tax rate.

參考答案（透過數據特徵指出問題）：There is a highly polarized income structure in Boyboyland. The annual medium income of top 10% family had increased while that of low 10% family had decreased from 2018 to 2020. The wealth gap has widenen to the highest in 2020. To sum up, the income inequality in Boyboyland become more serious.

考試技巧：考試不會指明考生需描述數據趨勢或變化，考生若有足夠時間，可做齊兩部分，以確保不會失分。

解釋好處（社會角度）

JRE試題一定涉獵政策分析，並會指定要求考生説明該政策的好處。考生可從不同角度分析政策利益，其中一個角度是從社會層面出發。

社會角度一般泛指醫療、教育和社會公平性三方面出發，以下特別提供相關英文字詞：

1. Health: life expectancy at birth, distribution of effect of public health

2. Education: years of schooling, expected years of schooling

3. Social Justice: social cohesion, importance of access to labour market, equity（公平）& equality（平等）

例題：

> In an effort to provide quality education to all eligible children, the Potpotiza Government introduced 12-year free education. Children can receive free education usually from the age of 6 until 18. Recently, the education sector and parents requested further improvements to the early childhood education, i.e. the provision of 15-year free education.

題目：You are the officer responsible for the educational reform in Potpotiza. Please draft a discussion paper for the Prime Minister's educational bureau, setting out the pros and cons of the launching 15-year free education.

參考答案（社會角度）：The launch of 15-year free education promotes equity and equality in Potpotiza. Children from grassroot families can enjoy longer years of schooling.

考試技巧：考生可以把社會角度中的不同層次關鍵字串連在一起，例如上文提及的 equity、equality（社會公平性）和 years of schooling（教育）互相緊扣。

解釋好處（經濟角度）

絕大部分考生都能從經濟角度指出某政策的好處，唯大部分考生未能利用可量度（measurable）的數據表達，失去不少分數。

經濟角度一般以下列四個層面，再配分數據量度：

- 勞工市場：unemployment rate、real wages
- 宏觀經濟：per capita gross domestic product、external trade value
- 物價指數：consumer price index、inflation rate
- 財富分佈（disparity between the rich and poor）：Gini-coefficient

例題：

Goldenium is a small economy which relies heavily on imports. Recently, several trade unions urge the government to tackle the high rate of unemployment in the country.

題目：You are the officer responsible for the economic policies in Goldenium. Please draft a discussion paper for the Prime Minister's economic bureau, setting out the pros and cons of the introduction of tariff on imported goods.

參考答案（經濟角度）：The introduction of tariff can raise the employment rate of local workers in Goldenium. First, people will shift to buy local goods as the price of imported commodities raise. Second, the tariff revenue received by the government can be allocated to implementation of training programme for unskilled workers and the set up of job centres.

考試技巧：考生可從兩方面解說關稅（tariff）與就業的關係，包括當地人民以本地貨物作為入口物品的替代品（substitutes），及關稅收入可用作加強輔助就業服務之用。

解釋好處（環境角度）

環保產業乃世界潮流所趨，考生在講述某政策如何促進環保前，必須指出原有問題，再配合政策功效連貫講解。

	來源	影響
Air pollution	vehicle emission	diseases, e.g. bronchitis
Noise pollution	factory	ear malfunction
Land pollution	large disposal	the use of land
Light pollution	business operation	sleeping quality

例題：

Road traffic congestion has been deteriorating, with a general decline in car journey speed across Boozopida. On some major traffic corridors in the capital city, cars only travel at around 10 km/hour during peak hours, a speed that is not much faster than an adult's average walking speed of 4 to 5 km/hour.

題目：You are the officer responsible for the transportation policies in Boozopida. Please draft a discussion paper for the Prime Minister's transport panel, setting out the pros and cons of the electronic road pricing scheme.

參考答案（環境角度）：Air pollution increases with heavy traffic flow and consequently higher exposure time of commuters. Nitrogen dioxide emitted from vehicles brings about an increased likelihood of respiratory problems such as coughing and bronchitis. Limiting traffic congestion can have positive benefits on road travel such as air pollution, greenhouse gas emission, noise, and visual intrusion and road accidents.

考試技巧：在解說空氣污染影響方面，考生可利用中學甚至大專的知識作答，例如二氧化氮（nitrogen dioxide）的影響；另外，一般考生只聯想到電子道路收費與空氣污染的關係，較高能力考生可進一步把交通擠塞聯繫到 noise and visual intrusion。

解釋好處(政治角度)

在中文卷中，考生可提及中央與香港的關係，作為政治角色的解說，但英文卷屬 Data-based response question，雖然個案提及的經濟體極似香港，但考生絕不能以中港情況作答，而應盡量利用題目資料，輔以下列的政治角度作答。

- Legitimacy（認同性）
- Credibility （可信性）
- Mandate（授權）
- Governance effectiveness（管治有效性）

例題：

> Tsuenia's political system is an executive-led system. For a legislation to take effect, it requires the assent of the President. While the president has the power to dissolve the senate under certain circumstances, the senate also has the power to impeach the president, and the president is required to resign if he twice refuses to sign a bill passed by senate.

題目：You are the officer responsible for the political reform in Tsuenia. Please draft a discussion paper for the President's cabinet, setting out the pros and cons of the implementation of the ministerial system.

參考答案（政治角度）：Without the ministerial system, the executive has difficult to maintain an executive-led style of governance without risking its political credibility. A ministerial system would be conducive to improving the accountability of the executive and the executive-legislature relationship. In addition, a system under which ministers are appointed to take on political responsibility and accountability could help achieve a relatively more stable government, as experienced by some western countries. In other words, implementation of the ministerial system brings about the institutionalized political support and higher degree of legitimacy.

考試技巧：例題改自較早年的JRE題目，近年題目較少直接問及政治部分。唯考生亦要注意，若題目政策涉及對政治的影響，必須使用上述政治概念詞作答，方可取得上品成績。

解釋好處（文化角度）

這是考生經常忽略考慮的角度，導致考卷被評定為「未能作多角度分析」。這不是考生的問題，雖然香港是個「文化沙漠」，但特區政府一直有意推廣香港文化產業，幾年前更考慮設文化局，所以JRE經常在考卷加入文化層面元素作測試。

- Culture diversity（文化多樣性）
- Culture preservation（文化傳承）

例題：

> Candiza is a small economic city in Asia. Nearby countries has been promoting students' exchange and internship schemes of various forms and themes overseas.

題目：You are the officer responsible for the tertiary level education in Candiza. Please draft a discussion paper for the Prime Minister's educational bureau, setting out the pros and cons of promotion of students' exchange and internship schemes.

參考答案（文化角度）：The scheme exposes young people in Candiza to the prevailing economic, social and cultural landscape at both international levels, and enhance their inclusiveness of different cultures through exchanges with overseas youth.

考試技巧：文化角度相關概念較少，避免專有名詞重覆使用，考生可多利用相近詞，例如以 inclusiveness of different cultures 來替代 culture diversity。

解釋好處（個人層面）

除了前文提及的五大宏觀角度外，考生亦可從不同持分者層面去分析政策的好處。JRE期望一般考生都有修讀大學的「微觀經濟學」入門，即能解釋一個政策對個人行為的影響，一般影響包括：

- Cognition（認知）
- Affection（情感）
- Behaviour（行為）

例題：

Lalapore's municipal waste had continued to grow over the last five years – and all the territory's landfills have been nearly exhausted. "Between 2015 and 2020, the amount increased at an average rate of 1.9 per cent per year, outpacing population growth of 0.8 per cent but slower than economic growth of 2.9 per cent", according to a research brief by a local university research.

題目：You are the officer responsible for the waste management in Lalapore. Please draft a discussion paper for the Prime Minister's environmental bureau, setting out the pros and cons of the implementation of a charging scheme for disposal of municipal waste.

參考答案（個人層面） The charging scheme enhances citizens' awareness about environmental protection and waste reduction. The use of economic incentive to drive behavioral change towards waste reduction and clean recycling can be effective as it imposts the disposal costs at the sources.

考試技巧：考生必須解說一個政策如何影響個人的行為。一般考生在作答有關垃圾徵費（如膠袋稅）的題目時，只能夠說把成本轉移到消費者身上，便可以利用經濟誘因來減少廢物數量，而忽略awareness（與cognition認知相近）及 behavior changes 等字眼。

解釋好處(家庭層面)

考生一般只會考慮政策對企業的影響，忽略了對家庭的影響。在現今香港提倡家庭價值的年代，考生須説明政策對家庭的影響，包括以下幾方面：

- Family structure（家庭結構）
- Family division of labour（家庭分工）
- Dual-income family（雙職父母）

例題：

> Dumbumdland is a western city where individualism prevails. Sons and daughters live separately with their father and mother.
> On the other hand, to alleviate the tax burden on salary earners, several political parties advise the Dumbdumbland government to consider announced various proposals of tax concessions in the financial year.

題目：You are the officer responsible for the taxation policies in Dumbdumbland. Please draft a discussion paper for the Prime Minister's economic bureau, setting out the pros and cons of the introduction increasing both the dependent parent & grandparent allowance.

參考答案（家庭層面）：Elderly households typically comprise retirees, who are living on their savings and/or financial assistance from the Government or other family members not living with them. The policy encourages the younger generation to take care of their parents by providing taxpayers with the dependent parent allowance under salaries tax and tax under personal assessment.

考試技巧：一般考生只會把增加免稅額與個人收入掛勾，但絕不能忽視對家庭結構改變，如增加父母/ 祖父母免稅額可鼓勵子女與長輩同居。

解釋好處（學校層面）

除了家庭層面外，學校層面亦是考生經常忽略的一個環節。一個政策非單單影響個人和家庭，更應該影響學校，這亦是現今特區政府的思維模式。例如衞生署於2016年推出「好心情@香港」，推廣正面人生觀，兩年後把計劃推展到學校，名為「好心情@學校」，鼓勵學生重視生命。換言之，JRE會觀察考生能否把政策轉化到學校層面，考生可試從以下兩方面入手：

- Educational level（教育水平）
- Curriculum（教育課程）

例題：

> Saburland government provides free-WiFi services covering venues including public rental housing estates, public hospitals, markets, parks, sitting-out areas, promenades, tourist spots, public transport interchanges and land boundary control points, etc. The government is considering to build Saburland as a WiFi connected city by extending free public WiFi services.

題目：You are the officer responsible for the technological development in Saburland. Please draft a discussion paper for the Prime Minister's Innovation and Technology bureau, setting out the pros and cons of the promoting Saburland as a WiFi-city.

參考答案（學校層面）：The setting up of WiFi-city helps establishing WiFi campus for all schools to facilitate e-learning through the use of mobile computer devices. For example, it enhances the supply of quality e-learning resources which can support specific and diversified learning needs in complementary to locally developed curriculum-based e-textbooks.

考試技巧：設立為 WiFi 城市第一當然促進旅遊業，但政府最忌就是只考慮商業利益，尤其現今特區經常被指責為官商勾結。所以考生在作答時，必須說明 WiFi 城市如何幫公營及私營機構。

解釋好處（社區層面）

考生經常混淆「社區」（community）和「社會」（society）兩個概念，一般人均以為兩者並無分別。但在特區政府角度來説，政策分析一般都包括該措施對社區的影響，尤其是涉及新界土地問題。政府轄下的民政事務局及地區辦事處，便是要主力在社區內落實各項政策。

例題：

> The economic growth of Parkaria is very good in recent years. However, as the development matures and with citziens becoming more aware of Parkaria's cityscape, the effect of high-rise and high-density development is more keenly felt.

題目：You are the officer responsible for the urban development policies in Parkaria. Please draft a discussion paper for the Prime Minister's urban development bureau, setting out the pros and cons of re-planning the old urban fabric to meet the current expectation.

參考答案（社區層面）：The policy can meet the growing community aspiration for a better living and working environment. The community is particularly concerned about the design, layout, massing, permeability and connectivity of development as well as provision of open space which directly affect both the quality and character of the surrounding area. The policy can help to achieve quality environment to meet the community aspiration.

考試技巧：一般考生只能簡單説明重建能夠讓居民享受較佳的生活環境，卻未有詳細説明當中的因由，考生可以嘗試利用一下：design、layout、massing、permeability and connectivity of development，甚至 open space 等專有名詞。

解釋好處（全民層面）

題目所指有時可能是國家，又或者一個城市。不論如何，政策其中一個好處就是加強該地的凝聚力，考生可參考以下四個有關用詞：

1. Social cohesion
2. Sense of belonging
3. Coexistence of identity
4. Harmony

例題：

Kennland is a crowded city which was founded in 1950s. The Urban Bureau is looking for and introduces enhancements where appropriate to ensure that its policies and procedures strike the right balance between being sufficiently responsive to the needs of people affected and sufficiently cost-effective to sustain a viable urban renewal programme.

題目：You are the officer responsible for the economic policies in Kennland. Please draft a discussion paper for the Prime Minister's urban bureau, setting out the pros and cons of the introduction of preservation arrangement in the urban renewal programme.

參考答案（全民層面）：The preservation arrangement recognizes the contributions to the local character, history and collective memory of the neighbourhood. Having a collective memory creates a sense of identification among residents of Kennland. People love Kennland which is their home and will therefore seek to work together for a better future.

考試技巧：不用多說，這是一般考生不能像到的層面，所以有些考生即使考了多次都未能過關，便是欠了這些層面性的分析。

解釋好處（國際層面）

這是大部分考生忽略的地方，亦正是現今政府需要具國際視野的人材。香港作為國際金融城市，實施政策舉足輕重，一個政務員/ 行政主任絕不能忽視政策對國際層面帶來的影響。

- Diplomatic relationship（外交關係）
- International image （國際形象）
- Hard power（硬實力）
- Soft power（軟實力）

例題：

Chocobia is an international finance and business hub. She achieved the top rank among other nearly economies on economic freedom and competitiveness in the past few years. However, she has lost its leading position in the international competitiveness rankings since last year.

題目：You are the officer responsible for the economic policies in Chocobia. Please draft a discussion paper for the Prime Minister's economic bureau, setting out the pros and cons of the developing dialogues with ranking institutes.

參考答案（國際層面）：Global rankings on economic freedom and competitiveness are of paramount importance to its international image and attractiveness to foreign investment. Not only would the dialogues allow the government to have a more comprehensive understanding on Chocobia's competitiveness in different aspects, the valuable researches by these international ranking institutes could also lend to the investigation into the strengths and weaknesses of Chocobia relative to nearby competitors, thereby assisting the Government's policy deliberation on augmenting Chocobia's leading position.

考試技巧：不要以為跟這些評分機構是無聊白痴的行為，城市大學正在做，香港政府亦有做。考生要留意一點：就算題目沒有直接問及 developing dialogues with ranking institutes，若考生遇上有關改善某經濟體的國際競爭能力時，亦可採納此論據。

解釋壞處（社會角度）

任何政策均有其利弊，政策的好處較容易辨別，考生只須考慮政策的目的，便可得知政策的好處。唯政策的壞處則較難分析，因為文章不會提及，考生需從多角度思考分析。

社會角度一般泛指醫療、教育和社會公平性三方面出發，以下將提供相關英文字詞：

1. Health - life expectancy at birth, distribution of effect of public health

2. Education - years of schooling, expected years of schooling

3. Social Justice - social cohesion, importance of access to labour market, equity（公平）& equality（平等）

例題：

For many years, Pigpenland has been viewed as a city of opportunities with ample opportunities for people to move up the social ladder through their own efforts. At present, the lack of new growth engines has restrained earnings growth and social mobility. The housing problem is the biggest conflict in Pigpenland. While rent becomes higher and higher, flats are smaller and smaller in area. Young people are unable to get married and establish their family due to the housing problem.
There have been calls for Pigpenland to adopt an effective strategy to diversify and restructure its economy by capitalizing on the global trend of the real estate development. This may help create more higher paid and higher skilled job opportunities, thereby enhancing earnings and occupational mobility.

題目：You are the officer responsible for the economic reform in Pigpenland. Please draft a discussion paper for the Prime Minister's economic bureau, setting out the pros and cons of boosting the local real estate industry.

參考答案（社會角度）：Pigpenland is riddled with conflicts of all kinds, giving a loose rein to the exploitation of real estate hegemony. The boosting of local real estate industry may not strike a balance amongst transportation, environment and conservation issues, thus giving rise to more social conflicts.

考試技巧：地產霸權（real estate hegemony）是香港現今重中之重的問題，但在各勢力（例如環境保育）持分者反對下，問題遲遲未能得到解決，考生亦應能詳細以英語闡述相關解釋。

解釋壞處（經濟角度）

大部分政策都有促進經濟發展的效能，所以不少考生都未能從經濟角度解釋政策的壞處。以下提供一些常用的考試答案：

1. 經濟成果未能與民分享：政策能推動 gross domestic product，但未能推高 per capita gross domestic product；

2. 開放經濟造成本地失業：external trade value 上升，可帶動該經濟體的物流業和商貿的同時，推高本地勞工的 unemployment rate；

3. 物價上漲普羅大眾未能受惠：工資名義的上升，並不代表實質工資（real wages）上升，這要視乎 inflation rate；

4. 經濟蓬勃不代表所有市民都滿意，要視乎 disparity between the rich and poor。

註：上述各項經濟學專有名詞，可翻查前章有關經濟角度的好處。

例題：

> Goldenium is a small economy which relies heavily on imports. Recently, several trade unions urge the government to tackle the high rate of unemployment in the country.

題目：You are the officer responsible for the economic policies in Goldenium. Please draft a discussion paper for the Prime Minister's economic bureau, setting out the pros and cons of the introduction of tariff on imported goods.

參考答案（經濟角度）：Putting tariff on imported goods discourages these products from being brought into Goldenium as it makes it more expensive. Essentially a trade war might happen when countries start imposing tariffs on each other's products.

考試技巧：貿易戰是近年大熱題目，考生必須熟讀。

解釋壞處(環境角度)

這個角度考生較易掌握，因為不少政策例如基建工程均會對環境帶來負面影響，唯部分考生在初中是以中文修讀地理科，未能準確用英語清楚闡述其論據。

	對人類的影響	對環境的影響
Air pollution	Chronic respiratory disease, lung cancer, heart disease	Acid rain, forest damages, global climate change
Noise pollution	Hearing impairment, hypertension, annoyance, and sleep disturbance	Altering the landscape of plants and trees, which depend on noise-affected animals to pollinate them and spread their seeds
Land pollution	Contaminated lands cause problems in the human respiratory system	Breed rodents like rats, mice and insects, who in turn transmit diseases
Light pollution	Disrupt circadian rhythms and contribute to sleeping disorders	Washes out starlight in the night sky and disrupts ecosystems

例題：

The Government and the franchised bus operators have started the research work for introducing the new long-haul bus services. The services will offer more spacious seating and all-seater service with more stops and more comprehensive passenger amenities in the bus compartment. Operators will not reduce their existing services on account of introduction of new long-haul bus services.

題目：You are the officer responsible for the environmental policies in Kayian. Please draft a discussion paper for the Prime Minister's environment protection bureaus, setting out the pros and cons of the long-haul bus services.

參考答案（環境角度）：The rise in energy consumption brings about the expansion of fuel consumption which in turn increases the corresponding emissions of pollutants and greenhouse gases.

考試技巧：考生必須分清楚不同影響的角度，例如塞車屬於社會角度，而空氣污染才屬於環境角度。考官亦期望考生能準確地運用不同角度作分析。

解釋壞處(政治角度)

香港特區政策一地兩檢、國歌法和「DQ 議員事件」均為其帶來負面的政治影響。唯在英文試卷中的 data 較少涉及上述複雜事件。一般而言，考生只須考慮下列四個角度，便可剖析政治影響：

- Legitimacy（認同性）
- Credibility （可信性）
- Mandate（授權）
- Governance effectiveness（管治有效性）

例題：

Tsuenia's political system is an executive-led system. For a legislation to take effect, it requires the assent of the President. While the president has the power to dissolve the senate under certain circumstances, the senate also has the power to impeach the president, and the president is required to resign if he twice refuses to sign a bill passed by senate.

題目：You are the officer responsible for the political reform in Tsuenia. Please draft a discussion paper for the President's cabinet, setting out the pros and cons of the implementation of the ministerial system.

參考答案（政治角度）：Ministers are accountable for the actions of their departments to the legislature and the need for a minister to resign if serious errors had been committed in his department should be implemented. This would impose barriers on the government to introduce important bills and lose credit for supporting thc government.

考試技巧：例題改自較早年份的 JRE 題目，近年 JRE 題目較少直接問及政治部分。唯考生亦要注意，若題目政策涉及對政治的影響，必須使用上述政治概念詞作答，方可取得上品成績。

解釋壞處(文化角度)

「文化多樣性」與「文化單一化」永遠都是評卷標準的指定答案。文化多樣性指一個地區可保存其獨有文化之餘，亦與西式的文化共存，例如香港的飲食文化便屬多元。另一方面，文化單一性是指當地文化未能得以保存，當地居民已轉向習慣西式文化，香港傳統手工業的衰落便是一個最佳的例證。

- Culture preservation（文化傳承）
- Culture homogeneity（文化單一性）

例題：

The economic growth of Parkaria is very good in recent years. However, as the development matures and with citziens becoming more aware of Parkaria's cityscape, the effect of high-rise and high-density development is more keenly felt.

題目：You are the officer responsible for the urban development policies in Parkaria. Please draft a discussion paper for the Prime Minister's urban development bureau, setting out the pros and cons of re-planning the old urban fabric to meet the current expectation.

參考答案（文化角度）：The re-planning the old urban fabric may bring about insurmountable negative impacts such as destruction of local economic activities or social and cultural characteristics if the area is to be redeveloped. In other words, the local socio-cultural characteristics and heritage buildings cannot be remained.

考試技巧：一個社區的文化不限於其建築，亦可以包括當地的經濟活動、生活模式和社區網絡等。

解釋壞處（個人層面）

在個人層面分析政策在個人層面帶來的壞處時，可從一些社會弱勢社群角度出發，包括 Ethnic minorities（少數族裔）、Grassroot families（草根階層）和 Disabled people（殘障人士）。

例題：

> Ngchia is a small country in the Asian Pacific region. She set a target of reducing municipal waste disposal rate by 40% on a per capita basis by 2022.

題目：You are the officer responsible for the waste management in Ngchia. Please draft a discussion paper for the Prime Minister's environmental bureau, setting out the pros and cons of the implementation of a charging scheme for disposal of municipal waste.

參考答案（個人層面） The scheme brings about huge financial impact on low-income families. The incur of extra expenses may lead to grassroots suffer from financial hardship.

考試技巧：弱勢社群關注個人物質生活質素，考生可多從政策對他們生活影響的角度出發，如收入、就業和居住環境等。

解釋壞處(家庭層面)

在對家庭影響的壞處方面，考生一般只從草根家庭角度出發，甚少從對家庭的影響出發。考生應考慮的是從以下三個「家庭角度」：

- Family structure（家庭結構）
- Family division of labour（家庭分工）
- Dual-income family（雙職父母）

例題：

> Family-friendly employment practices (FFEP) are measures voluntarily adopted by employers to help employees fulfil their work and family responsibilities simultaneously, thereby balancing their work and family lives. Generally speaking, FFEP may comprise special leave to meet employees' family needs, such as marriage leave, paternity leave, parental leave, filial leave and compassionate leave, etc.
>
> The Taitailand government expressed grave reservations about the adoption of FFEP by employers on their own accord. There was a view that FFEP should be cultivated through legislation.

題目：You are the officer responsible for the labour policies in Taitailand. Please draft a discussion paper for the Prime Minister's labour bureau, setting out the pros and cons of the FFEP legislation.

參考答案（家庭層面）：Legislating for FFEP entails various issues concerning policy, legal principle and implementation, e.g. birth within/ outside of marriage and proof of relationship between the employee. There may be concern about how the relationship between a male employee and a child can be easily and effectively ascertained if the child is delivered by a woman who is not the employee's legal spouse.

考試技巧：在現今單親世代，很多家庭崗位立法都需要著重上述細節。至於有關單親家庭的描述，可參考此段：Childbirth outside of marriage covers a number of different situations. For example, both parents of the child are single, or either one or both parents is/ are married but there is no legally recognised marital relationship between them.

解釋壞處（學校層面）

在現今特區政府和廣大市民著重下一代的發展下，考生一般都認為政策不會對教育界造成負面影響。非也，現今教育改革一塌糊塗，畢業生起薪點十年來不升反跌，考生正好從下列角度分析政策對教育層面的壞處。

例題：

> Saburland government provides free-WiFi services covering venues including public rental housing estates, public hospitals, markets, parks, sitting-out areas, promenades, tourist spots, public transport interchanges and land boundary control points, etc. The government is considering to build Saburland as a WiFi connected city by extending free public WiFi services.

題目：You are the officer responsible for the technological development in Saburland. Please draft a discussion paper for the Prime Minister's Innovation and Technology bureau, setting out the pros and cons of the promoting Saburland as a WiFi-city.

參考答案（學校層面）：On Bring-your-own-device, some teachers anticipated additional workload arising from the need to monitor students' proper use of their mobile computing devices in class and to get familiar with the operation of different devices. They expected clear guidelines and practical professional development programmes to be organised by the Innovation and Technology bureau. Also, low-income families might not be able to afford expensive devices for their children. Some parents may also concern about the impact on the health and eyesight of students.

考試技巧：設立為 WiFi 城市，第一當然促進旅遊業，但政府最忌就是只考慮商業利益，尤其現今特區經常被指責為官商勾結。所以學生在作答時，必須說明「WiFi 城市」如何幫助公營及私營機構。

解釋壞處（社區層面）

香港政府經常犧牲某些社區利益，以滿足整體香港的需求，例如將軍澳和屯門的堆填區、即將在大嶼山興建焚化爐等便是最佳例證。在 JRE 試卷中，考官著重要看的是考生能否分析政策對部分社區的不良影響。

例題：

The economic growth of Parkaria is very good in recent years. However, as the development matures and with citizens becoming more aware of Parkaria's cityscape, the effect of high-rise and high-density development is more keenly felt.

題目：You are the officer responsible for the urban development policies in Parkaria. Please draft a discussion paper for the Prime Minister's urban development bureau, setting out the pros and cons of re-planning the old urban fabric to meet the current expectation.

參考答案（社區層面）：The redevelopment work would destroy the character of the area, the original community of the localities concerned and was unfair to affected communities and parties, and some people even lost their jobs as a result. Affected shop owners might be willing to move, but they had difficulty in finding another suitable shop unit in the area even if they were willing to pay the costs. In relation to the Local Sports. Also, shop owners may concern about the unpredictability of the then prevailing market rental level.

考試技巧：一般考生只能簡單説明重建會影響當地居民生，未有從不同持分者角度出及，例如當地的小販。

解釋壞處（全民層面）

為什麼一個政策會對全民帶來壞處呢？這是未曾任職政府的考生較能想像和理解。筆者舉一個例子大家便懂：HKTV 發牌事件。

發牌予一個免費電視台，全港市民受惠。政府卻倒行逆施，為甚麼？當然背後是有一個較大的考慮，就是該電視台可能會帶出反政府的訊息。JRE要考的，就是要看看考生有沒有這樣宏觀思考的角度。

例題：

> Kennland is a crowded city which was founded in 1950s. The Urban Bureau is looking for and introduces enhancements where appropriate to ensure that its policies and procedures strike the right balance between being sufficiently responsive to the needs of people affected and sufficiently cost-effective to sustain a viable urban renewal programme.

題目：You are the officer responsible for the economic policies in Kennland. Please draft a discussion paper for the Prime Minister's urban bureau, setting out the pros and cons of the introduction of preservation arrangement in the urban renewal programme.

參考答案（全民層面）：In a crowded city, it is common to see the community had aspirations on reducing development intensity through lowering building height and plot ratio and the adequate provision of community facilities. There are conflicting demands because the introduction of preservation arrangement might hinder a reduction in the development intensity.

考試技巧：Preservation arrangement 的好處容易想到，但其壞處則麻煩得多，考生必須掌握對所有議題都懂得正反論證。

解釋壞處(國際層面)

在全球氣氛緊張的局勢下，政府經常會推出一些惠及本地利益而忽略對國際關係影響的政策，考生可從以下四個角度作論述：

- Diplomatic relationship（外交關係）
- International image （國際形象）
- Hard power（硬實力）
- Soft power（軟實力）

例題：

Pizzilia government is abided an Ordiance which provides that no person shall import, introduce from the sea, export, re-export or possess the endangered species of animals and plants, except under and in accordance with a licence issued in advance by the government. A total ban of local endangered species trade is considered necessary by the Administration for elimination of any potential front for illegal markets. The government would provide compensation to the owners of endangered species of animals and plants because they could still possess the above items for non-commercial purposes.

題目：You are the officer responsible for the economic policies in Pizzilia. Please draft a discussion paper for the Prime Minister's economic bureau, setting out the pros and cons of the total ban of endangered species.

參考答案（國際層面）：The provision of compensation in any form to the ivory trade may send a wrong message to lawbreakers that there is a prospect of compensation which may accelerate and/ or intensify the proliferation of the poaching of endangered species of animals and plants and stimulate smuggling of a large amount of illegal items into Pizzilia to launder with the legal stock for seeking compensation. In other words, It will merely complicate ban control and open up potential loopholes, which in turn will impede local law enforcement and confuse the public on the purpose of the trade ban. All these will not only significantly reduce the effectiveness of the ban, but also run contrary to the global efforts on conservation of endangered species of animals and plants and severely damage the international image of Pizzilia.

考試技巧：對國際層面影響的掌握程度要視乎考生的學歷背景和工作經驗。舉例，一般只有有意從紀律部隊任職轉考 AO/EO 的考生，才能想到 smuggling of a large amount of illegal items into Pizzilia to launder with the legal stock for seeking compensation。

解釋影響（僱主利益）

極多考生混淆好處、壞處與影響的不同。好處和壞處是要以不同宏觀角度及層面作分析，而影響則是主要針對社會不同持分者和既得利益團體。舉個例子：落實標準工時在宏觀經濟角度是損害香港的營商環境，而受影響的持分者則是僱主和僱員。反過來説，若在分析標準工時好處時，考生説好處是基層僱員受惠，那樣看來考生則未能考慮標準工時立法對整體營商環境的影響。

在分析政策對僱主影響時，一般考生只能提及 operating costs（營運成本），未能展示他/ 她的商業觸覺，考生可試用下列商業字眼：

1. Fixed cost（皮費）
2. Wages and salaries（工資與薪金）
3. Transaction cost（交易成本）
4. Rental cost（租金成本）

例題：

> There have been calls over the years from some sectors of the community, particularly the labour unions, to introduce standard working hours in Liligium to better protect the rights and interests of employees.

題目：You are the officer responsible for the economic policies in Liligium. Please draft a discussion paper for the Prime Minister's labour bureau, setting out the likely impact that the introduction of standard working hours would bring about to Liligium.

參考答案（僱主利益）：The implementation of standard working hours will undermine the flexibility of operation and increase the man-power cost of enterprises, particularly for the small-to-medium ones.

考試技巧：題目問的 impact 即是對各個持分者的影響，在僱主角度亦可再細分，例如標準工時立法對中小企帶來較大的影響。

解釋影響(僱員利益)

現屆政府極關注勞工權益，除了有意就標準工時方法外，亦正積極研究強積金對沖方案，所以考生必須充分顯示對勞工議題有充足認識。

在分析政策對僱員影響時，一般考生只能提及工資上升（標準工時立法）或享受較多休息時間（標準工時立法），忽略了其他勞工議題，包括：

1. Retirement protection
2. Working environment condition
3. Unemployment benefits
4. Re-training programmes

例題：

There have been calls over the years from some sectors of the community, particularly the labour unions, to introduce standard working hours in Liligium to better protect the rights and interests of employees.

題目：You are the officer responsible for the economic policies in Liligium. Please draft a discussion paper for the Prime Minister's labour bureau, setting out the likely impact that introduction of standard working hours would bring about to Liligium.

參考答案（僱員利益）：The long hours of work may have adverse effect on their physical and mental health as well as employees' family and social life. Some employees may not be compensated for their overtime work and their rights are not fully protected.

考試技巧：標準工時立法除了保障員工在充足的休息外，亦同時規劃了勞工合法需列明工作時數及超時工作的薪金等，加強對勞工的保障。

解釋影響(基層利益)

不要以為基層利益不得以政府重視，每年特區綜援出「雙糧」甚至「三糧」，大量非牟利組織亦每年均向政府請願，要求當局正視基層需要。在分析政策對基層影響時，考生可從改善物質生活著手。

1. Safety net（保護網）
2. Social security assistance
3. Housing facilities
4. Transportation allowance

例題：

> In recent years, the local community has been concerned about the impact of high private housing rental costs on low-income tenants. There are increasing suggestions for the Government to introduce tenancy control in Ngtszia.

題目：You are the officer responsible for the housing policies in Ngtszia. Please draft a discussion paper for the Prime Minister's home affairs bureau, setting out the likely impact that tenancy control would bring about to Ngtszia.

參考答案（基層利益）：With relatively limited bargaining power, low-income tenants might be forced to pay excessively high rent or evicted by landlords without "justifiable cause" if without tenancy control. Under the tenure control, landlords might not be allowed to circumvent rent control by evicting existing tenants and entering into new leases with new tenants willing to pay higher rent. Moreover, landlords are bound by tenure control to give "justifiable causes" to the sitting tenants when they decide not to continue with a tenancy.

考試技巧：租金管制的另一項影響是管制租金的上限，考生亦可加入答案內，豐富個人論述。

解釋影響(中產利益)

一般考生喜歡從非物質角度探討中產家庭需要，例如改善環境污染、擴大可用空間、提高教育質素等，卻忽略了香港中產同樣重視物質生活。舉例說，每年財政預算案影響社會上均出現派錢與退稅的爭議。

雖然個案中考核的不是香港的情況，但考官亦期望考生能明白中產對物質和非物質生活同樣重視，考生可採用下列專有名詞：

1. Tax rebate（退稅）
2. Tax reform
3. Environment living environment
4. Educational investment

例題：

MokSauland's home prices rose for a 30th straight month in December 2019, as apartments continued to sell for record amounts across the city. The government said that to meet the demand, the proportion of public housing of the new housing production should go beyond further.

題目：You are the officer responsible for the housing policies in MokSauland. Please draft a discussion paper for the Prime Minister's housing bureau, setting out the likely impact that the increase in the proportion of public housing would bring about to MokSauland.

參考答案（中產利益）：Those belonging to the lower stratum of the middle-class families wished to buy their own homes because they were not eligible for public housing, and were suffering from high private flat rentals. The suggestion to adjust the proportion of public housing might not be favourable to these families. To adopt a higher proportion of public housing would reduce the supply of private flats, which might in turn fuel the already high prices in the private residential market.

考試技巧：魚與熊掌，不可兼得。增加公屋比例，惠及基層利益同時卻損害中產利益。在闡釋論據時，亦應有深入見解，例如減少私人樓宇比例會進一步刺激私人市場樓價。

解釋影響（外資利益）

香港是一個國際金融城市，任何政策均會觸動外資的神經。考生必須能夠從外資角度思考問題，方能取得考官芳心。稅項及優惠當然是其中一項外資關注的事項，但其他因素如社會因素等亦不容忽視。2016年農曆年初，時任行政長官宣佈香港發生暴動事件，假期後股市即跌800多點，這便是最佳的例證。考生可利用下列字眼考慮外資的利益：

1. Tax rebate and subsidies
2. Political stability
3. Environmental pollution
4. Social atmospheres
5. Government commitment

例題：

There is a concern whether Lihkgery could stay competitive in attracting foreign direct investment to drive economic growth, spur innovation and technology, and create jobs. As a measure to attract inward investment, the government saw it important for the Administration to render sufficient aftercare support services to overseas investors, especially in the recruitment of professionals, specialist staff and local talents. To meet the pressing demand by foreign investors for sufficient international school places, the senate members suggested the Administration to give priority to non-local students in admission to international schools and increase the intake ratio of non-local students for new international schools.

題目：You are the officer responsible for the educational policies in Lihkgery. Please draft a discussion paper for the Prime Minister's education bureau, setting out out the likely impact that giving priority to non-local students in admission to international schools would bring about to Lihkgery.

參考答案（外資利益）：The policy might hurt the interest of local students and parents. It is that the government assign a dedicated team to assist overseas education institutions to set up international schools in Lihkgery and advise local international schools on expansion plans.

考試技巧：這是一種比較取巧的做法。就是透過説明政策的壞處，再提出建議如何補救這壞處。一般而言，我們都不建議考生採取上述做法，唯考生在一些論據上缺乏足夠英文字詞，這種取巧的方法是沒辦法之中的辦法。

解釋影響（公務員利益）

民建聯前主席譚耀宗被傳媒質問：「為什麼貴黨在政府絕大部分的議案都投贊成票？」他這樣回答：「如果方案得不到民建聯的支持，那政府一定不敢把方案拿出來。」同樣，一個方案若影響公務員的重大利益，那後果一定不堪設想，馬時亨的「仙股事件」便是最佳例證。

1. Civil servant's payment
2. Process of policy formulation and implementation
3. Staff morale
4. Job security
5. En/ dis-couragement
6. Bullying and harassment

例題：

In recent months, the prevailing economic downturn has prompted a suggestion from some Senate Members and others for a pay cut for civil servants as a political gesture. At the same time, concerns that civil service pay beyond the entry ranks for certain grades had fallen out of line with private sector pay have revived. In response, the Chrisaland government has firmly rejected the call for an arbitrary and out-of-cycle pay reduction for civil servants, reiterating that there is a long-established mechanism and timetable for civil service pay adjustments. But the Administration notes that it had been over a decade since we last conducted an overall review of the civil service pay policy and system and that there is a need to examine whether the current arrangements continue to meet present day circumstances.

題目：You are the officer responsible for the civil servants' compensation policy in Chrisaland. Please draft a discussion paper for the Prime Minister's civil servant bureau, setting out the pros and cons of the conducting an overall review of the civil service pay policy and system.

參考答案（公務員利益）：The review may lower staff morale, hinder team spirit and recognition, and discourage encouraged continuous improvement in the delivery of public services, particularly the civil service has been facing increasing pressure with growing public demand for government services in both quantity and quality.

考試技巧：這是一款較早期的 JRE 題目，近屆政府已甚少討論公務員凍薪或減薪方案。唯考生若遇到有關題目時，一定要從公務員角度出發和分析，如上例所示，因為閱卷員也是一名公務員呢！

倡議者利益(勞工團體)

這是比較冷門的題目，是分析政策倡議者的利益/ 看法/ 價值觀。一般考生只能簡單從倡議者的自身經濟利益出發去思考，忽略了他們的看法/ 價值觀。以勞工團體為例，他們所考慮的除了是工人的待遇外，尊嚴亦是他們重視的其中一項元素。

1. Wages and salaries
2. Labour law protection
3. Overtime-work payment
4. Fringe benefits
5. Work-life balances
6. Career prospects

例題：

Responding to some trade unions' concern about the Administration's lack of action in the promotion of collective bargaining through legislative means, employers' associations advised that it was not an appropriate time to introduce compulsory collective bargaining by legislation in Wingard as the community had no consensus on the issue. The employers' federation also advised that it was promoting the establishment of tripartite committees comprising representatives of employers, employees' organizations and the government.

題目：You are the officer responsible for the labour policies in Win-gard. Please draft a discussion paper for the Prime Minister's labour bureau, setting out the implications of the trade unions' requests.

參考答案（勞工團體）：The promotion of collective bargaining through legislative means further provides protection for employees by providing them with the right to claim civil remedies against their employers for unreasonable and unlawful dismissals, including dis-missals on ground of union discrimination.

考試技巧：很多考生混淆方案的好處和倡議者的利益。集體談判權的好處當然是保障工人利益，而倡議者（即勞工團體）的 Implication 是他們經常受到僱主不公平對待，包括 unreasonable and unlawful dis-missals, including dismissals on ground of union discrimination。

倡議者利益(資方團體)

資方團體最關心的一定是自身的利益，所以最低工資、標準工時、強積金對沖等都一概反對。在英文DBQ題目中，考生必須能充分利用英語講述政策對資本家的影響。

1. Operating costs
2. Profits tax
3. Tax allowance
4. Operating costs
5. Retained profits
6. Shareholders' interest

例題：

> Kakaopia is a small Asia country and adopts a competitive taxation system to promote economic development, while maintaining a simple tax regime. Business associations urge the government reduce corporate tax rates to attract more foreign businesses.

題目：You are the officer responsible for the economic policies in Kakaopia. Please draft a discussion paper for the Prime Minister's economic bureau, setting out the implications of the business associations' requests.

參考答案（資方團體）：The tax savings by enterprises can be reinvested in upgrading their hardware or software, providing and creating more employment and training opportunities thereby boosting their overall operation and efficiency which in turn could bring in additional profits in the future. In other words, the lower tax will benefit private enterprises by alleviating their tax burden. The tax saved will enable these enterprises to pursue their profits maximization objectives.

考試技巧：在這題目可再次看到政策的好處（pros）與倡議者利益（implications of the proponent's requests）的分別，好處必定是 It will gain international publicity mileage in promoting Kakaopia as a preferred investment destination. The lower tax rate will reduce the tax burden on enterprises, especially small and medium enterprises and startup enterprises. This will help foster a favourable business environment, drive economic growth and enhance Kakaopia's competitiveness。

唯商界口頭說減稅可以吸引外資，實際當然是為了自身利益，這便是政策的好處（pros）與倡議者利益（implications of the proponent's requests）的分別。

倡議者利益(租戶)

租戶一般可分為兩組，一組是公屋租戶，另一組是私樓租戶，包括割房或普通私人樓宇等等。不論是什麼形式的租房，所關注的一定離不開租金問題。其實港英政府曾經實行租務管制，而近年亦開始有聲音要求政府重新推出租務管制。有見及此，本書特設這一章講解相關字詞。

1. Rent control
2. Tenant
3. Property owner
4. Turnover rate
5. Maintenance
6. Key money

例題：

In recent years, the local community has been concerned about the impact of high private housing rental costs on low-income tenants. There are increasing suggestions for the Government to introduce tenancy control in Ngtszia.

題目：You are the officer responsible for the housing policies in Ngtszia. Please draft a discussion paper for the Prime Minister's home affairs bureau, setting out the implications of the proponent's requests.

參考答案（租戶）：Tenancy control can be justified when there is market failure, such as those stemming from imbalance/ unequal bargaining power between landlords and tenants.

考試技巧：租管的好處當然是 The intended effects of tenancy control are not limited to the provision of affordable rental housing and guarantee to sitting tenants of their rights to continue to rent the property after the expiry of the tenancy agreement.（詳細可返回前章「基層利益」參與。而倡議者利益則是要平衡租戶與業主之間的議價能力。）

倡議者利益（業主）

考生一般混淆業主和業主立案法團（稍後提及）的利益都是一致，實際當然有所不同，看看香港現今圍標的情況必能略知一二。業主最關心的不是能租出去的回報有多少，或樓宇升值的幅度，最重要的是物業稅（property tax）以及轉讓限制。

例題：

Property owners that keep their property vacant effectively withdraw supply of occupiable properties from the market, thereby increasing pressure on prices and rents in Lamziland. The government noticed that and has been studying vacant property tax. The tax will be a 1 per cent tax on the capital value of the taxable property. For example, if the taxable property has a capital value of $500,000, the applicable tax will be $5,000. The tax will only apply to the owner of a property that is unoccupied for more than six months within a year.

Homeowner's association demands a number of practical exemptions, recognising there are some legitimate reasons for a property being left vacant such as holiday home.

題目：You are the officer responsible for the housing policies in Lamziland. Please draft a discussion paper for the Prime Minister's home affairs bureau, setting out the implications of the proponent's requests.

參考答案（業主）：As the issue of high rentals could not be easily solved in the short run as the construction of new buildings would take time, the flat price would rise sharply lay with the demand-supply imbalance in housing. Maintaining certain vacancy rate of private residential flats would further raise the price and rent significantly.

考試技巧：這可能有點陰謀論，但在 implications 這一部分，合理地猜測 proponent 的背後動機是可得分的。

倡議者利益（業主立案法團）

題目未必每次都會談及業主立案法團，因公屋租戶是用「互助委員會」制度而非法團制度。貪污問題是近年在法團中最嚴重的問題，包括：

- Bid-rigging（投標操縱）
- Bribing（賄賂）
- Intimidating（恐嚇）

例題：

Privately owned multi-storey buildings are where most people live in Dayhidia, there is a general community consensus that owners have the responsibilities to properly manage and maintain their own properties. Building management involves various stakeholders, such as owners, tenants, owners' corporations or other forms of residents' associations, and property management companies. Lack of communication among stakeholders can easily lead to conflicts. Major owners' association urge the government to allow residential properties to set up Owners' Corporation to take responsibility for handling the above problems.

題目：You are the officer responsible for the housing policies in Dayhidia. Please draft a discussion paper for the Prime Minister's housing bureau,setting out the implications of the proponent's requests.

參考答案（業主立案法團）：Forming Owners' Corporations allows residents to appoint their own property management companies to undertake the management and maintenance works of their private properties. It also encourages owners' participation in properties management to facilitate owners' autonomy. This management duties include but not limited to awarding of contracts and management of Owners' Corporations' Fund.

考試技巧：這一道題目又再展現政策缺點和倡議者利益的分別。成立法團的壞處是容易出現貪污問題，而倡議者利益則不能寫他們希望貪污，只能從權力慾方面說明。貪污等問題的英文闡釋如下：Some lawbreakers may secure major maintenance contracts of private properties by means of bid-rigging, bribing and intimidating, and make huge profits from these works. The exorbitant maintenance costs will pose heavy financial burdens on many owners, and even lead to violent confrontations between owners and members of owners' corporations.

倡議者利益(中小企)

超過九成的香港企業屬中小企。中小企的定義較為繁複，服務性行業是指少於 50 個僱員，而非服務性行業（如工業）則指少於 100 名僱員。這些繁複的定義無須理會。最重要的是，中小企對成本特別感到壓力，任何勞工法例修訂均會反對，而同時會要求政府提供稅項優惠及津貼。

例題：

Chanszland government provides direct subsidy to SMEs for participation in export promotion activities. However, the subsidy has been criticized for being insufficient to business overseas missions, particularly placing advertisements in printed trade publications or trade websites targeting export markets. The SMEs associations urge the government increase the maximum amount of subsidy under the per SME.

題目：You are the officer responsible for the economic policies in Chanszland. Please draft a discussion paper for the Prime Minister's economic bureau, setting out the implications of the SMEs associations' requests.

參考答案（中小企）：The global financial turmoil has caused economic downturn in many countries. Many SMEs are facing a decrease in orders from major export markets. They need supporting measures to help them get new orders and develop alternative markets, as well as to alleviate some of the costs in carrying out promotion activities. The primary financial need of SMEs is to get loans for working capital. In the prevailing business climate, the raise of the maximum amount of subsidy enhance SMEs' performance by providing greater support and flexibility to SMEs.

考試技巧：題目指明 SMEs 出口津貼，考生可先指出現今出口業面對的問題，然後再指出倡議者利益。

倡議者利益(智庫)

智庫英文為「Think-tank」，「Tank」這個字即是智庫的意思，指集結超多高智慧的人倡議政策。其中一個最出名的政策倡議是大埔船灣填平用作建屋之用。所以在香港人心目中，這班智庫實際「智障」。但出奇的是，政府各部門撰寫大量文件，回覆指這建議不可行。

政府極重視智庫的倡議，唯後者的提案一向有自身的利益，包括：

1. Religious belief
2. Political party interest
3. Profit motive

例題：

The Government, being the largest employer in Jocktia, is a strong advocate for equal opportunities in employment. All candidates in an open recruitment exercise are assessed on the basis of their ability (including language proficiency). In order to maintain effective communication for delivering public service, the Government specifies appropriate English as part of the entry requirements for appointment.

To ensure that ethnic minorities, who do not speak English, have equal access to job opportunities in the Government, a think tank has been making on-going efforts to urge the authority setting lower language proficiency requirements.

題目：You are the officer responsible for the civil servant recruitment policies in Government. Please draft a discussion paper for the Prime Minister's civil services bureau, setting out the implications of the proponent's requests.

參考答案（智庫）：This requests provides an insight on the implementation of the measures to facilitate the employment of ethnic minorities in the civil service.

考試技巧：若遇上字數不足的情況，考生可以講解 setting lower language proficiency requirements 的例子，包括 requiring written proficiency in either English or their native language; identifying posts within the grades concerned for which lowered English language proficiency requirements would not compromise satisfactory performance of the relevant duties; replacing written test in English by group interview 等。

倡議者利益(宗教團體)

香港在道德議題的立法上，仍然非常保守。舉個例子，香港的強姦罪只適用於男性而非女性，在科技發達藥物倡明的新世代下，女性犯下強姦罪是絕對有可能，而外國亦有多宗案例。另一方面，香港有同性平權和婚姻法例仍非常落後，事關每次提案均被大批宗教組織反對。他們所關注的包括：

1. Moral standard
2. Ethical standard
3. Religious belief

例題：

Experiences of discrimination reported by the LGBTI (intersex-specific human rights issues) people were extensive, in the areas of employment, education, provision of services, disposal and management of premises, and government functions in Waiyiulza. Many LGBTI people saw legislating to protect them from discrimination as an important and necessary first step to protect their basic human rights.

On the other hand, number of religious group suggested that using public education alone as a strategy in eliminating discrimination on the grounds of sexual orientation and gender identity is adequate and effective.

題目：You are the officer responsible for the equal opportunities policies in Waiyiulza. Please draft a discussion paper for the Prime Minister's home affairs bureau, setting out the implications of the proponent's requests.

參考答案（宗教團體）：Religious groups concern about possible discrimination against them in the context of the possibility of introducing anti-discrimination legislation on the grounds of sexual orientation, gender identity and intersex status. particularly concerned that legislation could create a conflict with their rights such as freedom of expression, freedom of thought, conscience and religion, and the right to privacy. They advocate that developing forums, workshops and training sessions can increase dialogue and better understanding between different groups in society on issues relating to LGBTI equality.

考試技巧：高能力考生會利用 reverse discrimination（逆向歧視）去闡述宗教團體的關注。若考生對這議題未有深入認識，可參考上述答案，在教育層面上多作解釋。

倡議者利益(婦權團體)

一場「Metoo」的婦女運動，使女權再一次得到世界的關注。不論你怎樣看女權主義，婦女團體近年在香港確實是大有聲勢。她們所關注的包括：

1. Sexual orientation
2. Gender identity
3. Discrimination
4. Parental leaves

例題：

Goldenium A is a developed economy with a very high per capital gross domestic product. The labour force is an important asset to the development of Goldenium, and the Government accords top priority to improve labour welfare, the Prime Minister says.

題目：The Goldenium Prime Minister is reviewing the statutory maternity leave. To allow mothers more time to spend with and take care of their newborn babies, it is proposed to extend the statutory maternity leave from the current 14 weeks to 20 weeks.

參考答案（婦權團體）：Inadequate maternity leave forces new-born baby mothers detach from the labour marketand earn lower wages after resumption of work. On the other hand, supportive working environment and promotion of family-friendly employment policies boost female employment and fertility rate, especially in giving birth to subsequent children.

考試技巧：男性考生經常忽略了女性離開職場一段時間後，難以爭取一個較高的工資，這亦解釋了為何香港長期存在 gender pay gap。

倡議者利益(少數族裔)

每個地區都有少數族裔，香港亦不例外。不論考生對他們感覺如何，作為政府必須重視他們的需求，而同時平衡本地居民的利益。

1. Social integration
2. Translation and interpretation services
3. Long-term relationships and mutual trust with local residents

例題：

Goldenium A is a developed economy with a very high per capital gross domestic product. The labour force is an important asset to the development of Goldenium, and the Government accords top priority to improve labour welfare, the Prime Minister says.

題目：You are the officer responsible for the economic policies in Goldenium. Please draft a discussion paper for the Prime Minister's economic bureau, setting out the pros and cons of the introduction of tariff on imported goods.

參考答案（經濟角度）：The introduction of tariff can raise the employment rate of local workers in Goldenium. First, people will shift to buy local goods as the price of imported commodities raise. Second, the tariff revenue received by the government can be allocated to implementation of training programme for unskilled workers and the set up of job centers.

考試技巧：在國家層面中，考生亦可加插經濟角度及社會角度等元素，作更深入透徹的分析。

其他考慮因素(持分者參與)

一般考生臨場才想甚麼是其他考慮因素？其實就是題目未有提及的因素，本書將會列出十個八個題目未有講及的因素，供考生備用。

第一個就是持分者的參與程度。理論來説，持者分參與程度愈高，政策的制訂便能更為完善。唯近年特區政府舉辦的諮詢會少之又少，可見持分者參與的利弊，屬其他考慮因素之一。

例題：

The economic growth of Parkaria is very good in recent years. However, as the development matures and with citizens becoming more aware of Parkaria's cityscape, the effect of high-rise and high-density development is more keenly felt.

題目：You are the officer responsible for the urban development policies in Parkaria. Please draft a discussion paper for the Prime Minister's urban development bureau, setting out the pros and cons of the re-planning the old urban fabric to meet the current expectation, the likely impact it would bring, and all other relevant considerations.

參考答案（持分者參與）：The extent that the Parkaria government allow other stakeholders to be involved should be taken into consideration. The policy would be more feasible and legitimate as the planning would be more comprehensive after taking into different opinions in account. However, it is difficult for the society to reach a consensus, and may result in more arguments and conflicts among various stakeholders.

考試技巧：考生除了需正反說明利弊外，亦可指出及解釋政府應否讓持分者作更多參與。

其他考慮因素(群眾意見)

前文提及現屆政府甚少在政策制訂過程中聽取民意。當然，一個負責任和民主的政府，應該多與群眾交流意見，畢竟政府所用的錢屬於公帑；唯政府又有另一個思量：究竟收集到的意見是否真正對香港好？政府要做的是平衡各方利益，唯在收集公眾意見過程中，各持分者只會就自己的利益或角度發聲，若政府聽從他們的意見，可能耽誤了香港的長遠發展。

例題：

The economic growth of Parkaria is very good in recent years. However, as the development matures and with citizens becoming more aware of Parkaria's cityscape, the effect of high-rise and high-density development is more keenly felt.

題目：You are the officer responsible for the urban development policies in Parkaria. Please draft a discussion paper for the Prime Minister's urban development bureau, setting out the pros and cons of the re-planning the old urban fabric to meet the current expectation, the likely impact it would bring, and all other relevant considerations.

參考答案（群眾意見）：Since urban planning policy requires a large amount of public money, in the process of policy-making, a democratic and responsible government should listen to the opinions of the general public. However, various stakeholders may focus on their own interests and needs, neglecting the long term development of society.

考試技巧：考生經常把「持分者參與」和「群眾意見」混為一談，其實是兩個不同的論據。

其他考慮因素(少數權益)

群眾意見只是社會大多數人的意見，政策制訂最容易忽略的便是少數權益。以高鐵為例，原居民的利益屬於少數，仍容易被忽視。

例題：

The economic growth of Parkaria is very good in recent years. However, as the development matures and with citizens becoming more aware of Parkaria's cityscape, the effect of high-rise and high-density development is more keenly felt.

題目：You are the officer responsible for the urban development policies in Parkaria. Please draft a discussion paper for the Prime Minister's urban development bureau, setting out the pros and cons of the re-planning the old urban fabric to meet the current expectation, the likely impact it would bring, and all other relevant considerations.

參考答案（少數權益）：Public consultations take a long time and may result in delay of the projects, hindering the economic development and weakening the competitiveness of Parkaria. However, it is important to take the interests of different stakeholders into account by incorporating them in a multi-channel consultation process or consultation committees, so as to protect the rights of minority.

考試技巧：少數群體不會積極為自己的權益發聲，所以要透過不同方法搜集他們的意見，包括 multi-channel consultation process，又或者 consultation committees 等。

其他考慮因素(社會利益優先)

一般人均認為政府政策考慮以社會利益作大前提屬「阿媽係女人」的常識，但在一些多爭議的議題，例如社區重建中，政府會收到大量以個人利益為依歸的意見，考生此時必須説明政府是要以社會利益為大前提。

例題：

> The economic growth of Parkaria is very good in recent years. However, as the development matures and with citizens becoming more aware of Parkaria's cityscape, the effect of high-rise and high-density development is more keenly felt.

題目：You are the officer responsible for the urban development policies in Parkaria. Please draft a discussion paper for the Prime Minister's urban development bureau, setting out the pros and cons of the re-planning the old urban fabric to meet the current expectation, the likely impact it would bring, and all other relevant considerations.

參考答案（社會利益優先）：Urban planning affects the economic and social well-being of the whole community. Also, re-planning the old urban fabric requires high public expenditure and tax payers' money. Hence personal interest should not hinder the development of the society. The government should be respnsibile for safeguarding the interest of the whole. In other words, the needs of the Parkaria should take priority.

考試技巧：若考生的答案字數不足，可在社會利益優先這方面再加闡釋，例如社會利益與個人利益之間實無衝突。英文寫法如下：The interest of the whole community and personal interests are not mutually exclusive. Development in the society will benefit the individuals in the long run.

其他考慮因素（個人利益優先）

這是某所高等大學學府教的技巧，理論比較複雜。讀者若有不明白地方可重覆再看。這是現代新興的「左翼」（或稱「左膠」）理論，國際代表者是美國前總統 Barack Obama，他們主張政府應重視個人利益得失。

例題：

The economic growth of Parkaria is very good in recent years. However, as the development matures and with citizens becoming more aware of Parkaria's cityscape, the effect of high-rise and high-density development is more keenly felt.

題目：You are the officer responsible for the urban development policies in Parkaria. Please draft a discussion paper for the Prime Minister's urban development bureau, setting out the pros and cons of the re-planning the old urban fabric to meet the current expectation, the likely impact it would bring, and all other relevant considerations.

參考答案（個人利益優先）：Being a developed society, when formulating urban planning policies, which pose huge impact on individuals, the government should give prior consideration to the requests and views of these individual, consider alternatives to minimize or compensate for their loss. In other words, legitimate personal interest (e.g. the right of private ownership of property) should be given prior consideration in a civilized society like Parkaria.

考試技巧：坊間有不少應考JRE的課程均有教授上述技巧，唯近年局方有一句ban phrase，就像英文公開試的「one coin has two sides」這種句子不能使用，考生務必注意：serving the social interest may be manipulated by the government for the interest of the rulers。

其他考慮因素(兩者差異)

不少考生都對於「其他考慮因素」這個要求感到疑惑，都寫了政策的利弊以及影響，還有甚麼要作補充呢？這個其他考慮因素要求就像面試時考官問你有沒有其他問題一樣，是作補充之用。當然，這亦不可以亂寫，因為是計分的。

例題：

Candiza is a small economic city in Asia. Nearby countries has been promoting students' exchange and internship schemes of various forms and themes overseas.

題目：You are the officer responsible for the tertiary level education in Candiza. Please draft a discussion paper for the Prime Minister's educational bureau, setting out the pros and cons of promotion of students' exchange and internship schemes, the likely impact it would bring, and all other relevant considerations.

參考答案（兩者差異）：Working and interacting with nearby countrymen creates better understanding of Asia situation and fosters sharing of common values. However, the Candiza students may further consolidate a strong sense of local identity because of large political/ economic/ cultural differences between Candiza and other nearby Asia countries.

考試技巧：交流和實習計劃好處當然是擴闊考生視野，英文即為 global citizenship（全球公民），詳細答法可參考前文。在其他考慮因素這環節中，考生必可質疑這些計劃的作用，可能經過這些計劃後，Candiza 的人更加會認同自己的 local citizenship（本地居民身份）。

就如一些參與國內交流和實驗的香港學生一樣，回來以後不會增加其愛國情懷，反而認識到中港區分，更認同自己是香港人。

其他考慮因素(跨世代)

坊間教授答題技巧的其中一個方法是短中長線分析，即一個方案的影響分三個時間段作説明。以垃圾徵費為例，短期影響是加重中小企及基層負擔，中期影響是減少香港垃圾量，改善環境生活質素。

例題：

Candiza is a small economic city in Asia. Nearby countries has been promoting students' exchange and internship schemes of various forms and themes overseas.

題目：You are the officer responsible for the tertiary level education in Candiza. Please draft a discussion paper for the Prime Minister's educational bureau, setting out the pros and cons of promotion of students' exchange and internship schemes, the likely impact it would bring, and all other relevant considerations.

參考答案（跨世代）：Both material and spiritual life for future generations will be enhanced as a result of the consideration of the principle of inter-generation equity in the economic development strategy. It is because energy and environmental resources could be available for future economic development with a slower pace of economic growth rate and green economy. The overall economic development brings about more job opportunities and public resources for the improvement of the material and spiritual life of future generations continuously.

考試技巧：在可持續發展的 12 個原則中，以「跨世代」最重要，故答案應包括以下字眼：

1. Future generations 2. Inter-generation

3. Future development 4.Future generations

其他考慮因素（諮詢）

近年特區政府經常縮短諮詢時間，亦限制諮詢形式，諮詢是否愈多愈好？考生可考慮以下因素：

1. Responsiveness to public needs
2. Chance of power abuse by officials and mal-administration
3. Legitimacy of government policies
4. Transparencies of decision-making process
5. Communications between the general public (or among other stakeholders) and the government

例題：

Person-to-Person telemarketing calls (P2P calls) refer to telephone calls involving real person interactions used as a marketing tool by businesses/trades to promote goods or services to customers/potential customers. It is one of the most common modes of commercial promotion in Pinkybuda. As in some other jurisdictions, the rampant proliferation of P2P calls in recent years has caused nuisance to phone users who do not wish to receive any such calls or such calls at such high frequency. In response to requests for strengthening the regulation of P2P calls, the Pinkybuda Government is going to conduct a public consultation.

題目：You are the officer responsible for the establishment of a regulatory regime to strength the regulation of P2P calls. Please draft a discussion paper for the Prime Minister's Commerce and Economic Development Bureau, setting out the pros and cons of the establishment of a regulatory regime, the likely impact it would bring, and all other relevant considerations.

參考答案（諮詢）：The establishment of a regulatory regime does not proceed on the basis of a constructed consensus. Views from the business or related sectors and those from the public will be starkly dichotomised. Industry associations, trade practitioners and companies of the relevant trades are all not favour to the regime. They will express unanimous opposition to a legislative approach for regulating P2P calls. Members of the public including most political parties and district/community groups will be in clear support of strengthening the regulation of P2P calls by legislation. They will opine that self-regulation has not been effective and only through a statutory regime could the P2P call telemarketers be deterred from calling and causing nuisance to them.

考試技巧：諮詢雖然能採集不同意見，但是若意見有大分歧，難以使社會達成共識。

其他考慮因素(生活質素)

要衡量一個地區是否發達，不可以只觀看本地生產總值（Gross Domestic Product，簡稱 GDP），因為一些污染等指數沒有計算，之後經濟學家又發明了國民經濟福利（Measurable Economic Welfare，簡稱MEW），但其他學者又發現壽命和教育程度亦影響一個地區的發展程度，於是學者們又研發了「人類發展指數」（Human Development Index，簡稱 HDI）。

當然，上述這些字詞只有大學本科修讀相關課程才能知曉，但一般考生都懂得生活質素（Quality of life，簡稱 QOL），即從五方面分析一個政策的利弊。

例題：

> The economic growth of Parkaria is very good in recent years. However, as the development matures and with citziens becoming more aware of Parkaria's cityscape, the effect of high-rise and high-density development is more keenly felt.

題目：You are the officer responsible for the urban development policies in Parkaria. Please draft a discussion paper for the Prime Minister's urban development bureau, setting out the pros and cons of the re-planning the old urban fabric to meet the current expectation, the likely impact it would bring, and all other relevant considerations.

參考答案（生活質素）：It could cause the closure of some business and loss of jobs. Without new urban reform, small and medium scale enterprises may operate with lower costs of production and sustain their business even in times of economic downturns. It could result in more stable salary for some employees, with a corresponding stability in their standard of living.

考試技巧：對於某些群體（例如勞工），穩定性比其他影響生活質素的元素更為重要。

衡量準則(影響的規模)

考卷經常給予兩個選項，要求考生衡量哪個選項較好，這正正是考核學生能否舉出相關的衡量準則（Parameters），最簡單準則的當然是受影響人數/ 規模，例如某項政策的影響是屬於整個社會性或是地區性等。

例題：

In view of the growing consumption of organic food in Two-republic and to facilitate the further development of the organic food sector, the government is considering how to enhance consumer education and information about organic food. The following are two options reconmmended by a local University:
(i) raising consumers' awareness, including dissemination of information to organic producers, traders and consumers
(ii) promoting organic certification schemes

題目：You are the officer responsible for the urban development policies in Two-republic. Please draft a discussion paper for the Prime Minister's health bureau. Which options do you think government should go with? Explain your answer with justifications. Apart from the options listed, you may suggest new policy initiatives.

參考答案（影響的規模）：Considering option (i), whether the consumers are able to differentiate authentic organic food from the conventional/ counterfeit food is questionable. Option (ii), therefore, is a more practical alternative might be for the industry to step up the policing efforts.

考試技巧：選項 (i) 根本毫無影響，反而選項 (ii) 能促使生產商或進口商多做功夫！

衡量準則(影響的相對程度)

「相對」(relative)這個概念是一般考生較難掌握的地方。考生可換一個角度思考,即政策的效果能否互相抵銷。舉例說,開發生態旅遊景點便能抵銷旅遊業對環境保育造成的破壞。

例題:

Patroance Government commissioned a team of economists, physicians, epidemiologists, and public health specialists to conduct a study of health care system. According to the Report, the long-term financial sustainability of the current health care system was highly questionable.
The team put forward the following options for discussion:
(i) raises users' fees for hospital care and each day of hospitalisation
(ii) compulsory enrolment in an insurance to cover large medical expenses of the population. Employers and employees together would pay 2% of wages

題目:You are the officer responsible for the health-care policies in Patroance. Please draft a discussion paper for the Prime minster's health bureau, which options do you think government should go with? Explain your answer with justifications. Apart from the options listed, you may suggest new policy initiatives.

參考答案（影響的相對程度）：In view of the increasing demands which the ageing population, advances in medical technology and rising expectations for quality health services would put on the public health budget, option (ii) is more effective in reducing the burden on the next generation and strengthening the long-term financial sustainability of the public health care system.

考試技巧：既然日後的醫療費用愈來愈貴，選項 (ii) 便能抵銷這問題，反之選項 (i) 則未能做到。

衡量準則(影響的相對重要性)

一個影響是否重要呢？以環境保育為例，香港政府為了保護元朗濕地，西鐵站便未能按原意接駁至落馬洲站。另一邊廂，國內則相對不著重環境保育，以經濟發展凌駕環境保育。

例題：

To tackle the emerging problem of urban decay in Koritica, the Government carry out urban renewal projects. The members of urban development bureau consider the following compensation policies:
(i) monetary compensation: the rate of compensation based on the value of a seven-year-old notional flat
(ii) non-monetary compensation: e.g. "flat-for-flat" and "shop-for-shop"

題目：You are the officer responsible for the urban development policies in Koritica. Please draft a discussion paper for the Prime Minister's urban development bureau, which options do you think government should go with? Explain your answer with justifications. Apart from the options listed, you may suggest new policy initiatives.

參考答案（影響的相對重要性）：There are practical problems in option (ii). The preference of the affected parties on the location and configuration of the replacement units might be difficult to satisfy. Holding sufficient housing stock for yet-to-be affected residents was another problem that needed addressing. On the other hand, option (i) was already generous and flexible.

考試技巧：選項 (i) 只是影響政府財政，但選項 (ii) 需政府儲備一大堆樓房作樓換樓之用，其相對負面影響當然較大。

建議新政策(嶄新概念)

一些務必取得過關的考生都會寫新建議，務求令閱卷員眼前為之一亮。的確，若新建議寫得好的話，英文 140/200 分是「走唔甩」的！但是否真的有能力去寫，卻又是另一個問題。本書會介紹兩個最基本寫建議的方法。

例題：

To tackle the emerging problem of urban decay in Koritica, the Government carry out urban renewal projects. The members of urban development bureau consider the following compensation policies:

(i) monetary compensation: the rate of compensation based on the value of a seven-year-old notional flat

(ii) non-monetary compensation: e.g. "flat-for-flat" and "shop-for-shop"

題目：You are the officer responsible for the urban development policies in Koritica. Please draft a discussion paper for the Prime Minister's urban development bureau, which options do you think government should go with? Explain your answer with justifications. Apart from the options listed, you may suggest new policy initiatives.

參考答案（嶄新概念）：New option of joint development between Koritica government and affected owners should be examined. A "people-oriented" approach to urban renewal should be provided to those who wished to participate in joint development to do so. The feasibility of joint development would depend on the timeframe of the project, overall planning for the district and interest of affected owners.

考試技巧：選項 (i) 是賠錢，選項 (ii) 是樓換樓、舖換舖。假如考生真的要 100% 保證自己文章夠分取得面試資格，便可嘗試提出這種「非收錢、非收樓」，而是政府市民「齊發展、齊收錢、齊起樓」的雙贏模式。

建議新政策（2 合為 1）

「相對」（relative）這個概念是一般考生難以掌握的地方。考生可換一個角度思考，即政策的效果能否互相抵銷。舉例說，開發生態旅遊景點便能抵銷旅遊業對環境保育造成的破壞。

例題：

Patroance Government commissioned a team of economists, physicians, epidemiologists, and public health specialists to conduct a study of health care system. According to the Report, the long-term financial sustainability of the current health care system was highly questionable.

The team put forward the following options for discussion:

(i) raises users' fees for hospital care and each day of hospitalisation

(ii) compulsory enrolment in an insurance to cover large medical expenses of the population. Employers and employees together would pay 2% of wages

題目：You are the officer responsible for the health-care policies in Patroance. Please draft a discussion paper for the Prime Minister's health bureau, which options do you think government should go with? Explain your answer with justifications. Apart from the options listed, you may suggest new policy initiatives.

參考答案（2合為1）：The Patroance Government can enhance the public-private interface with incentive schemes to leverage private participation so that those with means are encouraged to use private healthcare services. The government can launch pilot schemes to facilitate the sharing of patient information with private practitioners, provision of information on private providers to patients to facilitate their choice, and public-private collaborative service models. In the long run, promoting private public collaboration can eventually achieve a long term sustainability of the healthcare system.

考試技巧：既然選項 (ii) 是繼續由政府介入醫療事務，而選項 (i) 是以市場機制處理醫療負荷問題，倒不如來一個公私營合壁的方法。

CHAPTER 2

JRE 文法技巧

文法重點 (1)：簡潔有力

坊間一些 JRE 班，會刻意教授一些累贅字眼（redundancy），這其實只會模糊考官的改卷形式。

筆者現綜合多年來的「考官內部報告」，列出部分常見累贅寫法和改善方案。

例子(1)：垃圾徵費

錯誤示範：The government had better take measures to do something for trash reduction now.

正確寫法：The government had better recycle and sort trash for trash reduction now.

例子(2)：選擇方案

錯誤示範：It goes without saying that option (1) is the best policy.

正確寫法：Undoubtedly, option (1) is the best policy.

坊間常見錯誤例子改正：

1. due to the fact that 應改成：because

2. during the time that 應改成：when

3. in today's world 應改成：today

4. in spite of the fact that 應改成：although / though

5. has the ability to 應改成：can

6. in an efficient way 應改成：efficiently

文法重點 (2)：主動語氣

坊間的導師指JRE考試中必須爭分奪秒。句子以被動句來寫不但語氣弱，而且字數會被拖長，所以他們均建議考生盡量選擇主動語氣來寫。筆者再於下列補充，有以下情形，考生應以被動句去撰寫：

1. 特殊句型

(a) It is reported that ... (d) It is said that ...

(b) It is thought that ... (e) It is announced that ...

(c) It is believed that ... (f) It is expected that ...

2. 強調動詞、受詞，而非執行者

主題：垃圾徵費

Residents lack a sense of public moral. Bags of trash were thrown on the roadside.

3. 行為者顯而易見

主題：打擊吸煙

New laws against smoking should be proposed and formulated.

文法重點 (3)：第三人稱

千萬不要以第一或第二人稱作答JRE試卷，「必肥」無疑。多用第三身稱，例如：

1. People
2. They
3. Those who
4. Almost everyone
5. Anyone
6. One
7. No one

主題：駕駛法例

錯誤示範：If you hit and run, you will face a harsher punishment.

正確例子：If people hit and run, they will face a harsher punishment.

若果考生英文能力稍遜，可多用 it，詳情可觀看前頁第一章。

文法重點 (4)：句型

香港大學生一定懂得以下五個基本句型，不懂的話別考了，一定不合格。

1. S+Vi（I agree）

2. S+Vi+SC（It sounds great）

3. S+Vt+O（It hurts me）

4. S+Vt+IO+DO（She made me a cake）

5. S+Vt+O+OC（I heard Carrie mention her proposal）

除了以上五個基本句型外，考生必須展現能撰寫其他句型的能力。

1. 複句

主題：交通規例

例句：Not following the traffic rules results in many road accidents.

註：考生可利用phrase或clause作字句開始，但評分只會計算作同一句型——複句，所以句子結構要多加變化

2. 同位詞

主題：房屋政策

例句：Cold and hungry, the homeless issues must be tacked immediately.

3. 反問句

主題：生物多樣化

例句：Shouldn't we care more about our earth and try to live peacefully with other creatures?

4. 倒裝句

主題：交通規例

例句：Hardly had the drivers yielded the road when they noticed an ambulance behind.

其他常用倒裝句字詞包括：never, rarely, seldom, no sooner, not until, in no way, by no means, under no circumstances…

文法重點 (5)：引述數據

JRE 英文試卷為 Data-based question，考生須引用題目資料作答。一般考生只懂寫 from source A, from source B……句式毫無變化，效果甚差。

以下提供一些變化字句作字句開首參考：

1. Considering ...
2. Concerning ...
3. Providing that ...
4. Judging from
5. In view of
6. Regarding ...
7. Provided that ...
8. Given that ...
9. According to
10. Based on

除字句開首外，考生可用下列字句加插句中，以引述資料的作用：

1. ... suggest that ...
2. ... have ... to do with ...

若要引用多一個資料，可使用下列一些對等詞，包括：

1. Both ... and ...
2. Not only ... but also ...
3. So ... that ...; so that ...
4. Neither ... nor ...
5. Either ... or ...

文法重點 (6)：段落結構

考官一般只需靠 Topic sentence 便能猜到考生的實力。實力較弱的會在開首寫：First, the feasibility of proposal (i) is higher，這些句子未能帶出整個段落的重點，必被評為中下品甚或更差。

Topic sentence 之後便是段落的發展，多用轉承詞（transitions）和連接詞（conjunctions）。

常用轉承詞（transitions）

1. 表示「列點」：first, second, third ... last
2. 表示「此外」：moreover, furthermore, whats' more
3. 表示「轉折」：however, nevertheless, nonetheless
4. 表示「因此」：consequently, therefore, accordingly
5. 表示「舉例」：for example, for instance
6. 表示「同樣地」：similarly, in the same way, likewise
7. 表示「轉折」：but, although, in spite of
8. 表示「目的」： in order that, so that, so as to

常用連接詞（conjunctions）

1. 表示「原因」：because, seeing that, since, as, in that, for

2. 表示「諸如此類」：and so forth, and so on

CHAPTER 3 中文試卷剖析

簡介

中文題目為 Essay-type question，考官看的是 based on your own knowledge，即考生知道多少。有些考生會先做中文卷，因為貪方便、我手寫我心。然而就筆者觀察，一般考生是先完成英文試卷，原因請翻看本書的後記。

本章 50 多篇不志在教大家寫議論文的格式，而是「貼題」！雖然未必能夠準確貼中題目，但本書提供的資料絕對是政府內部近年經常討論的政策，非政務官和行政官能參透。

考生可在考試前一個月，才開始溫習中文的各大議題，只要作答時配合適量的延伸，中文試卷必能過關。

設題方向

政府制訂政策時，必先考慮外國例子。在比較政策時，考生須考慮政府從中的角色及牽涉的持分者。

比較香港與英國的賭博條例，香港政府是保障香港賽馬會為「獨家唯一受注人」，所以馬會每年須向政府繳交大量賭博税，但同時因缺乏競爭者，馬會的賠率缺競爭性，導致外國莊家問題持續多年。

至於英國則容許多間博彩商互相競爭，賠率自然吸引，受注項目亦較香港多，包括籃球和選舉結果等，使更多市民容易沉迷賭博。

題目一般問考生，香港應否效法英國的賭博條例，考生要分析：

一、政府的角色是要提高更多的賭博收益（贊成）？還是要抑制賭博（反對）呢？

二、開放賭博競爭後會如何影響持分者利益？會否使更多人沉迷賭博？又或是即使不開放，香港市民會投注外圍，不受保障？

上述便是一般的設題方向，考生在閱讀本書時，可從中圈出各熱門議題的政府角色及對持分者的影響。

說明優點

一般題目都會要求考生說明建議政策的三個優點。一般而言，考生可從以下角度考慮：

1. 對社會、經濟、政治、文化和環境帶來的正面影響；
2. 對個人、家庭、學校、社區及整個地區帶來的正面影響；
3. 相對現行政策的好處/ 改善現行政策的不足。

政府即將會推行垃圾徵費，我們試利用上述三個要點，去分析這政策的三個優點：

a. 垃圾徵費減少固體廢物數目，改善土地污染問題；

b. 垃圾徵費改善周遭居住環境，尤其是接近堆田區的將軍澳；

c. 垃圾徵費提供直接的經濟影響，能有效改變市民處理垃圾的行為，包括鼓勵回收等。

當然垃圾徵費有其他好處，舉例其收費得益可用作回饋社會，但考生只能選三項優點作答，所以你們可依照上面三點回答，作多角度而深入分析。

說明缺點

一般題目都會要求考生説明建議政策的三個缺點。一般而言，考生可從以下角度考慮：

1. 對社會、經濟、政治、文化和環境帶來的負面影響；

2. 對個人、家庭、學校、社區及整個地區帶來的負面影響；

3. 相對於現行政策的壞處/ 未能替代現行政策的原因。

香港政府有意指出小學教科書全面電子化，我們試利用上述三個要點，去分析這政策的三個缺點：

a. 中小型書商難以轉型，使香港壟斷情況更為嚴峻；

b. 貧困家庭未能負擔電子教學的成本，包括購買平板電腦、教材和應用程式等；

c. 大量教師未有接受足夠培訓，例如如何確保學生拿著平板電腦時，不會「打機」或上社交網站，阻礙提昇教學質素等。

當然小學教科書全面電子化有其他壞處，舉例學生眼睛容易受損導至加深近視，但考生只能選三項缺點作答，所以你們可依照上面三點回答，作多角度而深入分析。

決定立場

在英文 JRE 試卷中，考生需解釋選擇哪個方案背後的 justification，只有言之成理便可，但在中文 JRE 試卷中，考生並不需要詳解其 justification，反之要留意題目要求。

一、若考生贊成題目建議，需闡述兩項詳細執行方案細節

二、若考生反對題目建議，需另提出改善現有情況的兩個方法

換言之，考生的立場其實是決定的自己的題型，若考生的立場是贊成，便會回答政策執行層面。若考生的立場是反對，便會回答政策建議題。

選哪個題型較好，沒一定準則，可看此書的教學和臨場形勢，再作決定。

執行方案

若考生同意立場，便要回答兩個執行方案細節，考生可從政策有效性和可行性去闡釋。

1. 有效性

有效性一般是指政策能影響更大的持分者，或政策的作用能夠長遠。舉例説，歧視條例應以廣義方式立法，包括同性戀者、異性戀者甚至變性者。而同時要配合恰切的宣傳和教育，方能成事。

2. 可行性

要促使新的歧視條例順利立法，最主要的當然是取得共識，否則會像「網絡二十三條」一樣被拉倒。若香港政府決定就性別取向歧視立法，有效性便是指政府如何有效訂定相關細節，包括舉行諮詢會和與各個主要持分者洽談，凝聚共識。

新建議

若考生不同意試卷立場，便要提出兩個建議改善現有狀況，一般建議包括：

1. 立法
2. 教育/ 宣傳
3. 加強執法/ 監管
4. 經濟誘因

若題目是有關立法，而考生反對的話，便是三個建議選兩個，例如：

a. 加強執法/ 監管
b. 經濟誘因
c. 教育/ 宣傳

一般常見的錯誤是，題目是有關加強教育，考生反對後又建議加強宣傳，那一定是不合格的，考生應只從立法、加強執法/ 監管、經濟誘因作三選二。

在撰寫建議時，考生除了要闡述建議能如何改善現況外，亦要補充建議方案中的預期困難和解決辦法，例如加強教育/ 宣傳的局限是需時長，未能立即見效，解決方案便是配合學校課程和社區活動，以及利用明星和社會名流作宣傳，加強影響力度等。

格式分佈

大家數一數，要分別寫政策的三個優點和缺點，之後加兩個方案執行細節（立場贊成）或兩個建議方案（立場反對），總共八點，連同前文後理，時間該如何分配呢？

先談內容分配，不少考生因內容重覆，因而導致「肥佬」。大家想想看，若考生立場是贊成的話，優點便容易與執行細節重覆；若考生立場是反對的話，缺點便容易與建議方案討論範疇相似。總之考官就要看到八點不同的論據。考生只要善用本書提及的論述方法，配合 50 多個熱門議題，相信不會犯此等錯誤。

至於時間方面，一般考生把時間除八，即八個論點，每個論點的平均時間相若，這是可取但不是最好。最好應該是三個優點和缺點時間佔 60%，後面兩個論點佔 40%，讓考官看看您能否撰寫好的執行方案/ 建議。那必定無失。

高官廢話實錄

全世界官場都有一種話叫「官腔」，好聽就是說了等於沒說，不好聽的就是廢話。大家都知道香港高官最多廢話，但這些廢話絕不能寫在JRE考試中。

高官在甚麼時候會説廢話？就是被記者問到稿件中沒有答案的問題，便「e e er er」。你作為負責寫預設答案的官員，怎能把這些廢話寫進去呢？換句話說，這些廢話為 ban phrase，切勿使用，現節錄如下：

1. 這只是一個個別事件……
2. 政府必定採取務實態度……
3. 我們不能排除這個可能性……
4. 這不是一個簡單的問題，而是結構性的問題……
5. 我們製訂政策上要取得平衡……
6. 我們要保持審慎樂觀的態度
7. 對於假設性問題，我們不會回應……
8. 若這項政策不能被落實，香港將被邊緣化……
9. 這個項目需循序漸進……
10. 這是歷史遺下來的問題……

兒童及家庭支援

為協助內地新來港定居人士適應本港的生活，政府提供多項支援措拖，包括在社區服務、教育、職業訓練及再培訓、社會福利服務、就業、公共房屋及公共醫療等範疇。

港人的內地妻子在批出單程證之前，輪候期間多以雙程證不時穿梭內地與香港照顧家庭和團聚。內地妻子只能短時間與家人團聚，而且不能在香港工作，只可以依賴丈夫的收入生活。此外，她們離港期間可能難於尋找別人照顧留港子女的日常生活。

同時近年內地孕婦在港分娩的數字不斷上升，有內地父母將子女留港，並交由親友照顧。而該等兒童以個人名義或由香港監護人代其申請綜援。

究竟政府如何妥善安排足夠資源予新來港家庭，同時又要避免資源被濫用？這是一個兩難之處。

兒童權利

由於性罪行可能會對受害人（尤其兒童）構成長遠影響，社會上一直存在訴求，希望減低前性罪犯從事兒童相關工作的風險，以及容許僱主查閱有關準僱員的性罪行定罪紀錄。

由 2011 年 12 月起，應徵與兒童相關工作的人士可以向警方申請，向其準僱主披露其性罪行紀錄。然而，近年數宗涉及從事與兒童或精神上無行為能力人士有關工作的人士的性侵犯事件經傳媒廣泛報道後，再度引起社會對查核機制的設計及運作的關注。關注的重點包括查核機制所涵蓋的工作類別過於狹窄，沒有包括志願工作者、私人家庭補習導師及已受聘的現職僱員。

另一邊廂，英格蘭及威爾斯全面推行兒童性侵犯者披露計劃，容許兒童照顧者申請查核個別人士是否有干犯兒童性罪行的紀錄，包括性罪犯名冊、刑事紀錄查核和兒童性侵犯者披露計劃。

公務員及公職人員

政府使用外判服務日增，帶來了不同方面的關注，例如外判工人權益保障及外判服務的質素。受聘於政府合約承辦商的非技術工人可能往往只能領取法定最低工資，而且缺乏加薪機會。

參照外國經驗，政府有以下兩個方案處理上述問題：

1. 採用「價格與質素方法」

新加坡政府的招標評審中，建議價格與質素的比例為 30:70，同時亦規定政府部門須提醒投標者將工資增長計入其投標價格，藉此推廣良好勞工管理做法。

2. 規管外判工人的薪酬

加拿大多倫多的「公平工資」、美國為聯邦服務合約工人實施的「專設最低工資」，以及新加坡的「漸進式薪金模式」（當中包括每年發放花紅）都是規管外判工人薪酬的例子。

工商業

中國目前是美國的最大貿易夥伴，而美國是中國的第二大貿易夥伴。由於香港是細小開放型經濟體，亦是轉口貿易樞紐，美國對中國進口商品加徵關稅，將會影響本港的轉口貿易。

除了本地出口商外，社會亦憂慮在內地從事製造業務的香港企業會受到影響。由於關稅增加會令製造成本上漲，預計部分企業為了避開這方面的影響，或會把業務遷往生產成本較低的其他亞洲國家。

有意見認為若香港善用大灣區的機遇，或可抵禦可能出現的震盪。又例如「一帶一路」倡議將有利企業開拓更多元化和新興的市場，為香港經濟發展增添動力。

政制事務

在 2016 年 9 月舉行的立法會選舉，部分事項引起公眾關注，例如個別投票站在投票時間結束後仍有大量市民排隊等候投票，以及部分投票站的累積投票人數與所發出的選票數目不符。針對該等關注事項，選舉管理委員會建議選舉流程電子化應該是未來發展的方向，當中包括電子投票。

電子投票有四大存在優點，包括增加投票的準確度、加快點票工作、增加投票的簡易程度和減低行政成本。

唯電子投票亦有一定的潛在風險，包括數據操控及選舉舞弊的風險、長者未有具備足夠的資訊科技能力透過電子裝置自行投票等。

消費者保障

儘管《商品說明條例》沒有訂明須就預繳式服務合約設立強制性冷靜期，某些行業或商戶仍有為其貨品或服務提供冷靜期。例如美容業、保險服務業及電訊業，亦已各自引入「冷靜期」安排。

關於香港就消費合約設立強制性冷靜期的建議，這需考量一些基本問題，例如適用的消費合約種類、冷靜期內消費者可否享用貨品或服務、如消費者在冷靜期內享用了部分貨品或服務，而後來又提出取消交易的要求，退款如何計算、消費者如何行使取消合約權，以及退款的安排等。

發展

面對土地短缺，社會上有意見認為可考慮改變現有未發展土地的用途，以擴大已建設土地的面積。這些未發展土地包括林地、灌叢、草地或濕地。另外，香港現有郊野公園和受法定保護的特別地區，面積高達相等於全港土地面積約 40%。

土地供應專責小組於 2018 年 4 月 26 日，就土地供應來源提出共 18 項建議作公眾諮詢，包括四項短中期、六項中長期和八項概念性選項。短中期選項包括：發展棕地、利用私人新界農地儲備、發展私人遊樂場地契的用地（例如發展粉嶺高爾夫球會），以及重置或整合康文署的康體設施用地。

中長期選項包括維港以外近岸填海、發展東大嶼都會、利用岩洞及地下空間、發展更多新界發展區、發展內河碼頭、發展郊野公園邊陲地帶兩個試點（即大欖及水泉澳）。

具爭議性的概念性選項包括增加鄉村式發展地帶發展密度、在現有運輸基礎設施上作上蓋發展、利用公用事業設施用地的發展潛力、葵青貨櫃碼頭上蓋發展，以及填平部份船灣淡水湖作新市鎮發展。

西九龍文娛藝術區

一幅位於西九龍填海區南部面積達 40 公頃的海濱用地，將會發展成為世界級綜合文娛藝術區。

有關注西九文化區最終或會成為地產項目，而且該區的消費可能很高，令市民大為卻步。又，有關注西九文化區應是一項文化投資，並非經濟投資項目。就此，政府應採取發展本地文化和加強藝術教育等互補措施，以強化西九文化區計劃。西九文化區亦可強化和善用本港地緣文化優勢，並關顧少數族裔和本土特色文化。

此外，因興建廣深港高速鐵路西九龍總站引起的工程問題和延誤風波，阻礙當局在西九龍區妥當推行交通改善措施，改善周邊道路的暢達性。

濫用藥物

近年青少年吸食毒品的情況有上升趨勢，氯胺酮、大麻及搖頭丸是最普遍被濫用的三種精神藥物。大部分濫用藥物的學生從來沒有因濫用藥物問題向他人求助。

禁毒處曾建議在本地學校推行自願校本毒品測試，但這產生了一定的問題，包括私隱、可能產生的標籤效應、計劃的經費、所需的支援和轉介服務等。

另一方面，警務處已增設「警務處學校聯絡主任」的職位，以加強警方、學校、社工和社區之間的聯繫，協助預防教育，對高危學生及早作出支援，以至跟進毒品個案，但成效存疑。

經濟發展

本地的工作人口中，約有十分之三具備大學學歷。在經濟轉型過程中，開創高端職位的進度緩慢，故此大學畢業生最終從事較低技術職位的比例持續增加。

同時，經濟增長減速，對年輕世代的事業發展路徑和社會流動，均有一定影響。即使位處收入最高端的百分之十的年輕世代精英畢業生，他們即使能覓得高端職位，其每月工作收入仍然遠低於較早世代的同齡前輩。

雖然政府實施若干協助初創科技企業獲得財務資助的措施，但 34 歲及以下的年輕人在本地僱主中的比重，更在 1991 年至今由 25% 縮減至少於 10% 。參考外國例子，它們的創業研究更聚焦在創造寬鬆的規管環境及公平的競爭環境，支持所有經濟領域的可持續初創企業，而非傾斜於個別經濟界別的活動。

教育

「STEM教育」為科學（Science）、科技（Technology）、工程（Engineering）及數學（Mathematics）的縮略詞，它於 1990 年代發源於美國。有別於傳統的科目，STEM教育着重於跨學科知識、解決難題及創新技能的應用。在四個STEM科目中，數學是香港整個六年中學教育的唯一必修科目。綜合科學在初中仍為必修科目，但高中有多達 51% 學生沒有報考任何科學科目。

參考外國愛沙尼亞政府STEM教育，全國課程重點教授科學及數學：愛沙尼亞的所有學生均須修讀科學和數學，直至年滿 18 至 19 歲。

此外，科學和數學必修課程佔三年高中教育總課業量約三分之一。學校並會舉辦課外活動，以補充及延展課堂學習。該國為那些對STEM有濃厚興趣的學生設立專科高中學校，提供必修課程之餘，還會分配更多時間及資源予STEM科目。

選舉

「愛國者治港」的意思就是，回歸祖國後的香港要由愛國者治理，香港特別行政區的政權要掌握在愛國者手中。

1984年，鄧小平闡述了「一國兩制」方針的主要內容，其中著重談了「愛國者治港」問題。他說：「凡是中華兒女，不管穿什麼服裝，不管是什麼立場，起碼都有中華民族的自豪感。香港人也是有這種民族自豪感的。港人治港有個界線和標準，就是必須由以愛國者為主體的港人來治理香港，未來香港特區政府的主要成分是愛國者，當然也要容納別的人，還可以聘請外國人當顧問。」

為了全面落實「愛國者治港」原則，推進「一國兩制」實踐行穩致遠，香港完善了其選舉制度，當中包括

一、重新構建選舉委員會、訂定委員的產生辦法、投票人和候選人資格;

二、訂定行政長官的產生辦法及相關事宜;

三、更新立法會的組成及產生辦法、訂定選民資格等。

電力、能源及食水供應

香港供水來自收集雨水、輸入東江水及抽取海水用作沖廁。今天，香港享有穩定的食水供應，但不能就此將水視作理所當然的資源。

以下列出兩項值得留意的範疇：

第一，雖然用戶無需就使用海水沖廁付費，但沖廁供水可能因水管滲漏及非法取水等原因而流失的問題。香港須減少沖廁供水的流失量，以節省就提供海水及淡水沖廁所招致的開支。

第二，在香港，用水效益標籤計劃及使用節水器具均屬自願參與的節約用水措施。但新加坡及澳洲規定強制參與用水效益標籤計劃。

環境事務

處置固體廢物是本港廢物管理的其中一個主要挑戰。家居及工商業的廢物約佔所有固體廢物約七成，而當中按類別分析，廚餘是都市固體廢物中最大部分，佔總量約四成。

另一方面，從本地最後一間廢紙回收再造廠關閉後，現時所有收集所得的廢紙皆出口至其他地方回收再造，當中以內地為主，但內地最近已收緊廢物進口規定。

參考外國例子，垃圾徵費制度不單對強制性的家居廢物按量收費計劃及強制性的家居廢物分類發揮補充作用，亦為循環再造業提供所需財政支援。

少數族裔

香港的巴基斯坦、印度及尼泊爾人士，面對較高的貧窮風險，部分原因在於這些族裔家庭的在職成員人數較少，但須撫養的子女數目卻較多。

此外，他們的教育水平較低，從事低技術工作，亦影響了他們的每月家庭收入。

政府現行支援措施問題如下：

1. 就業：少數族裔人士職位配對的成功率偏低，反映由勞工處提供的各項就業服務均未能有效協助少數族裔人士持續就業。

2. 教育：對少數族裔學生在中文科考試成績未如理想，他們難以在社會向上流動。

3. 福利：關愛基金設有多個援助項目，但可能由於宣傳不足，少數族裔人士的受助比率偏低。

金融財經服務

全球致力對抗氣候變化及其他環境問題，令不少綠色項目在未來蓄勢待發，而這些項目均需要投入龐大的資金。然而，目前香港在發展綠色金融方面沒有針對性策略。

建議有三點：

第一，擔當為綠色企業籌集股本資金的平台，把香港定位為籌集綠色資本的樞紐；

第二，透過保險將環境風險或相關項目風險轉移，使綠色保險的發展前景正面；

第三，訂立綠色金融產品的指引或標準以應對市場對「漂綠」的關注，例如設立符合國際標準的綠色債券認證計劃，此舉可增強投資者對資金運用的信心和吸引更多綠色產品發行者及投資者前來香港。

食物安全

政府自2010年7月1日起以強制性質推行預先包裝食物的營養資料標籤制度，讓消費者可以依據營養資料選擇食物。

然而，如何在包裝正面的標籤標示營養資料，目前並無劃一的做法。這使到營養成份列表上的資料複雜紛繁，很多消費者都覺得難以理解。

在英國、新加坡及澳洲，當地政府已積極制訂劃一的標示方式，使業界在食品包裝正面提供營養標籤時有所依循。政府同時致力加深公眾對該等營養標籤的認識，並推動市民在選購包裝食物時多加利用這些標籤，務求令該等營養標籤計劃更具成效。

政府及政治制度

目前，各大政黨大多根據《公司條例》註冊為公司。有建議香港特區成立「政黨法」，向政黨提供資助進行競選活動，或履行立法會或區議會工作。

有指若政黨完全依賴私人資助，可能會出現政治上不平等的問題。換言之，代表弱勢社群的政黨可能無法與代表商界利益的政黨在平等的情況下作出競爭。

亦有建議提出透過政黨法訂立適當規定，以增加政黨的財政透明度。然而，部分社會人士擔心，制定政黨法只會令政府有藉口訂立更多法律管制。

粵港澳大灣區

粵港澳大灣區（大灣區）是指由廣州、佛山、肇慶、深圳、東莞、惠州、珠海、中山、江門市和香港、澳門兩個特別行政區組成的城市群。大灣區的主要目標之一是提高區內合作，包括確定大灣區內各個城市的核心競爭優勢，並探索相互之間的互補合作模式。

雖然大灣區建設帶來許多機遇，但在推進的過程面對不少挑戰。廣州、深圳及香港三個主要城市，各自在人口、經濟規模方面相若，協同發展可能引發「龍頭之爭」。

又，大灣區仍存有粵港澳三套行政制度及三個地區稅制的差異，可能阻礙區內人流、物流和資金流的自由流通。部分香港專業領域例如法律、會計等未有直接的專業資格相互認證，需要參加內地相關專業考試以取得內地執業資格，然而，即使考取了內地專業資格亦未必能完全執業，例如香港公民取得內地律師資格，只能參與內地非訴訟法律事務。

再者，目前香港居民在內地工作如超過183日，須繳納個人所得稅。由於內地稅率較香港高，造成不少港人在內地工作時要計算逗留日數，以免遭內地政府徵稅。

醫療服務

近年輪候公營醫療服務的人數上升，加上嚴重醫療事故增多，公眾對醫院管理局提供的醫療服務滿意度已不如前。

民政事務

在香港，創業對很多人而言，一直都是一項挑戰，其中尤以青年人為然。政府早幾年前推出「青年發展基金」，協助年青人創業，但成效一般。

房屋

房屋供應緊絀，住宅租金不斷上升，加上公屋輪候時間頗長，迫使許多市民選擇居於面積較小的單位，導致分間樓宇單位（俗稱「劏房」）的需求居高不下。

除了居住空間狹窄外，消防安全、環境衞生、滲水及電線鋪設凌亂等問題，亦造成「劏房」的居住環境惡劣。

基於上述的情況，一些非牟利機構已開展計劃，與私人物業業主合作，由後者提供閒置單位，經裝修後以可負擔的租金把單位租予低收入家庭，作為過渡性房屋，但成效存疑。

人權

香港並無審定難民身份的制度處理庇護聲請。政府僅依靠專員公署處理此聲請。部分非政府機構表達關注，專員公署未能為在港的尋求庇護者提足夠保護，包括程序缺乏保證透明、獨立上訴、法律援助及司法覆核等程序上不公平公正等。

在香港，申請人不會獲准工作。在澳洲，工作許可會發給持有過橋簽證的申請人，這亦是另一項的關注問題。

資訊服務及通訊

隨着互聯網及流動技術蓬勃發展，全球各地政府紛紛透過電子方式（電子政府），市民提供公共服務和讓公眾參與活動。

香港電子政府發展的步伐遜色於其他領先地區，包括：

一、雖然市民可以從政府網站下載各式各樣的公共服務申請表格，但未必一定可把填妥的表格以電子方式交回。

二、香港市民雖然可在網上查閱諮詢文件，並以電郵方式提交意見，但政府並無設立一個中央電子平台，達至更有效的雙向溝通。

創新、科技及知識產權

行政長官在 2016 年施政報告中宣布，創新及科技局將與科研及公私營機構共同研究建設智能城市。研究範圍包括在巴士站和商場等地方提供免費 Wi-Fi 服務。

參考外國智能交通例子，巴塞隆拿市民可以使用特定的流動應用程式，查看各個地點的出租單車供應情況。除了推動單車使用外，巴塞隆拿全市亦設有 300 個供電動車輛使用的免費充電站。

司法制度

香港定罪率偏高，超過九成，高於英格蘭及威爾斯巡迴刑事法院的8成、加拿大法院以及澳洲法院的7成。

不同學者有不同解釋，包括：

一、是律政司只對理據充分的案件提出檢控，成功入罪率自然偏高。只有理據充分的案件才被檢控，定罪率偏高是可以預期的；

二、辯護律師經驗不足所致；

三、香港的司法制度仍然十分著重口頭證據，香港地方小，證人較容易出庭作供，但在其他司法管轄區證人要長途跋涉出庭，所以定罪率相對偏低；

四、香港政府在刑事訴訟法律援助方面的開支偏低，刑事抗辯資源匱乏。

諮詢及法定組織

政府一向倚重多個不同的諮詢及法定組織，就其政策提供意見和提供服務。現時政府有約 500 個諮詢及法定組織。

政府委任諮詢及法定組織非官方成員的原則是「用人唯才」，當中考慮有關人士的才幹、專長、經驗、誠信和參與服務社會的熱誠。

在新選舉制度下，選舉委員會增至一千五百人，其中一百五十六席將由分區委員會、地區撲滅罪行委員會和地區防火委員會的委員互選產生，所佔的席位佔整個選委會超過一成。由民政事務總署署長委任的地區諮詢委員會，主要包括70個分區委員會、18個地區撲滅罪行委員會及18個地區防火委員會。

民政是政府的「眼」及「耳」，需要經常聆聽地區的聲音，為政策把脈。作為政府的地區諮詢委員會，分區委員會、地區防火委員會和地區撲滅罪行委員會，一直是政府多元廣泛諮詢途徑的重要一員，也是民政長期的好夥伴、好拍檔，大大小小的地區政策都要靠他們幫忙才能「落地」。

漁農業

本章分為三個部分，包括為漁民而設的貸款基金、檢討停止簽發海魚養殖牌照的措施及水耕種植。

為漁民而設的貸款基金

基金由政府於 1960 年設立。宗旨提供貸款讓漁民轉而從事可持續發展的漁業。業界要求政府對漁業發展貸款基金作注資，以便進行遠洋捕魚，政府對此意見表示保留，解釋是相較其他界別，政府為漁業提供的財政資助，已是較為優厚。

檢討停止簽發海魚養殖牌照

政府自 1987 年起實施停止簽發新海魚養殖牌照的政策，以減低海魚養殖對海洋環境可能造成的影響。業界建議政府檢討停止簽發新的海魚養殖牌照，以推動香港水產養殖業的可持續發展。

水耕種植

隨著香港轉型為以服務業為主的經濟體系，加上耕地匱乏的緣故，本地農業活動正逐漸萎縮。水耕是一種以礦物質營養液在水中栽種植物的技術。由於水耕種植無需陽光及土壤，因此可作為室內或室外土耕以外的另一種耕種方法。

動物政策

本章分為三個部分，包括毀滅寵物的情況、制訂動物友善政策和規管動物火葬場。

毀滅寵物的情況

在香港，漁護署捕獲流浪動物後，會將有關動物出售或供市民領養。約 90% 的被遺棄的動物（包括流浪動物及棄養動物）在指定時間內 未能售賣或被人收養，最終被毀滅。

制訂動物友善政策

多個團體關注到本港維護動物福利的相關法例不合時宜，未能有效保護受虐動物，建議要求警方設立「動物警察」專責調查動物被虐待案件，警方回應指警區的刑事調查隊已有所需的人手、能力及經驗處理有關案件，有需要時會考慮調派指定隊伍調查。這種安排令警方可靈活調配資源，較成立「動物警察」更為有效。

規管動物火葬場

目前政府並無專門規管動物善終服務（包括火化服務提供者）運作的牌照制度。由於部分動物火化服務提供者位於工業大廈，有市民關注到該等服務提供者或會對鄰近　居民帶來衞生及空氣污染問題，例如動物焚化爐排放出煙霧及氣味。

人力政策及勞工事宜

目前，在持續進修基金下每名申請人一生可獲發還的資助上限為 20,000 港元。僱員日益需要終身學習新知識和技能，但本港的持續進修參與率近年相對偏低，這令政府關注到基金的資助金額是否不足，令有意進修者卻步，以及認可課程的範圍是否過於狹窄，許多課程都不被資助。

新加坡近年推行「技能創前程」計劃，為確保國民能夠負擔培訓費用，許多核准課程由新加坡政府大力資助支持，而僱主亦可獲得相等於僱員基本時薪 80% 至 95% 的缺勤薪金補貼。

藝術及文化發展

本章分為兩部分，分別是香港公共圖書館及香港電影業面對的挑戰。

公共圖書館

香港公共圖書館受到不斷擴張的互聯網的挑戰，根據調查，一半香港市民在過去一年並沒有使用公共圖書館的服務。當中原因包括館館藏量低於標準水平和電子書發展緩慢等。

電影業

電影業是香港六大創意產業之一，唯它卻面對著兩大挑戰：

第一，內地電影製作企業在電影製作上越趨成熟，亦可自給自足，這或會減少香港電影製作企業參與合拍影片的機會。

第二，香港電影需要面對外國電影的激烈競爭，尤以荷里活電影為然。

殘疾人士

目前，政府透過受資助的康樂界非政府機構為聽障人士提供多項社區支援服務，包括手語翻譯服務及訓練課程。

然而，聽障人士所使用的手語有不同的版本。學者把此情況歸咎於政府沒有為聽障人士提供標準化手語的正規教育。有建議指政府應把在香港所使用的手語標準化，並採用標準化手語作為為聽障兒童而設的幼稚園的輔助教學工具，以推動更廣泛地使用手語。

另一方面，有指政府應推動安老院舍和長者服務中心的醫護人員和照顧員學習手語，使他們能更好地與有聽障的服務使用者溝通，以提供適切的照顧和服務。

基本法

香港曾經歷五次釋法，分別如下：

第一次釋法：1999年

終審法院指所有香港永久居民在中國內地所生子女，不論出生時父或母是否已經是香港居民，全部有居港權。同年 6 月人大常委對《基本法》作解釋，指出只有獲批單程證的香港永久居民在內地所生子女才享有居港權，推翻終院裁決。

第二次釋法：2004年

人大常委就《基本法》中，修改特首及立法會產生方式作解釋，加入特首要為是否需要進行修改向人大常委提出報告，以及由人大常委依照《基本法》規定予以確定。

第三次釋法：2005年

2005 年 3 月特首董建華辭職，社會對補選特首任期有爭議。人大常委開會通過，補選特首任期為前任特首餘下任期。

第四次釋法：2011年

2008年剛果民主共和國與美國 FG 公司、中國中鐵在港有債務民事訴訟，由於案件涉及外國政府，而《基本法》第 19 條規定國防及外交等國家行為由中央政府負責。人大常委指特區須對剛果民主共和國實施絕對外交豁免權。

第五次釋法：2016年

2016 年 10 月，青年新政梁頌恒、游蕙禎在宣誓為立法會議員時宣揚港獨及辱華，立會主席梁君彥原本安排他們再宣誓。人大委員長主動提出就公職人員宣誓要求釋法，並於同年 11 月由人大常委會通過。

人口政策

香港人口急速老化，同時年長人士當中罹患多種慢性疾病和出現身體機能衰退的情況日趨普遍。社會上一直有意見要求政府增加資助安老院舍宿位數目，以應付日益增加的需求。

透過優化社區照顧服務可有助推遲甚至減少長者入住安老院舍，然而，香港並沒有訂立積極老齡化的指引政策。香港仍有很多年長人士在社區中積極生活，身體仍相當健康。這些體健長者應居於一個友待長者的環境，讓他們積極渡過晚年。

其他亞洲經濟體亦有推廣積極老齡化政策，讓長者保持健康、活躍和獨立。最近，新加坡近年啟動了「幸福老齡化行動計劃」，涵蓋12個範疇，分別為健康與保健、學習、義務工作、就業、住屋、交通、公共空間、尊重和社會包容、安穩的退休生活、醫藥和樂齡護理、為弱勢長者提供保障，以及針對老齡化的研究。

貧窮

香港的貧窮率有上升趨勢，同時香港的貧富懸殊情況愈趨嚴重，低收入人士面對經濟困境，影響了身心健康及家人關係；失業者常感有心理和社會壓力，導致家庭糾紛；同時，新來港的低收入家庭生活困苦，社交生活也受到影響。政府的教育、房屋及福利政策未能配合，影響貧窮兒童的成長，更可能因而令他們被困於貧窮循環中，難以脫貧。

舉教育方面為例，現時學校發展多元智能及全方位學習教學，學生不能像以往只熟讀書本便可以達到要求，而是要運用資源作多方面學習；例如，學生需要平板電腦上網或智能手機做功課，交費出外參觀等。可是，學校的基金未能全面資助所有年級的學生，而且政策不一，令貧苦學生再勤力也難以克服這些困難。

廣播事務及傳播媒體

香港電台是本港唯一的公共廣播機構。港台是政府部門，由廣播處長掌管。港台的經費來自政府撥款，並透過工商及科技局局長向政府負責。

港台服務質素一直備受關注。2005年時任行政長官的曾蔭權大力推動港台改革，包括停止轉播賽馬和舉辦中文金曲頒獎禮，理由是港台是公共廣播機構，不適 宜播放賽馬和娛樂節目。

同年時任政務司司長許仕仁表示，港台編輯主權獨立須維護，但港台用的是公帑，作為公營廣播機構，不是不可批評政府，但要多元化。

唯時任港台製作人員工會主席表示，政府高官發言違反港台的編輯自主及獨立。

私隱及個人資料保障

人對人促銷電話數目不斷上升，以致越來越多人對這些來電反感。此情況引起社會關注人對人促銷電話應否亦受到規管，例如設立「拒收電話登記冊」。商界反對並指強制規管人對人促銷電話可能會妨礙中小型企業的正常電子促銷活動，以及影響約過萬名從事有關行業的僱員的生計。

許多海外地方均設立「拒收電話登記冊」。在英國，法例規定所有進行促銷電話的機構，必須先將其電話撥打名單與「拒收電話登記服務」所登記的電話號碼及其機構內部的拒收來電名單進行比對篩選。撥打人對人促銷電話的促銷員亦須在通話中報上姓名，並須按對方的要求提供聯絡資料，違反規定可處罰款最高 50 萬英鎊。

儘管如此，政策不能完全杜絕當地的非應邀促銷電話：

首先，以市場調查名義撥打的電話不受規管；

第二，來電接收者通常無法提供人對人促銷電話來電者的確實身份，造成執法上的困難。

公共財政

社會上有意見認為政府在財政預算方面偏向保守，甚至刻意低估收入及高估支出，造成政府財政不穩健的假象，避免市民有太大期望，但這會影響政府的施政和資源分配能力，令市民對政府的信任度下降。

不少學者亦認為財政儲備水平並不是愈多愈好，政府要適當運用財政儲備，改善民生和促進經濟發展。的確，基層市民未能受惠於豐厚的財政盈餘和儲備，政府推出的短期紓困措施，未能徹底解決香港的貧窮問題。

然而政府對上述意見表示保留，解釋指政府回應，釐定充足的財政儲備是一項判斷，考慮因素包括儲備須提供足夠的資源，以應付未有撥備的負債，以及由於經濟周期回落、突發事件或社會結構轉變，帶來的財政壓力。當前的外圍經濟形勢仍然不穩定，政府須做好準備，務求維持香港的良好經濟基調和發展勢頭。

建造及建築

2010 年，九龍馬頭圍道一幢樓齡 55 年的唐樓突然全幢倒塌，造成四死兩傷。

成立業主立案法團是監察樓宇安全的其中一個可行辦法，但為舊樓成立法團其實存在很多不同的難處。例如業主資料散佚不全，很多業主立案法團委員均不具備檢驗維修質素等專業知識，不知如何監管維修質量問題。而業主立案法團本身亦往往出現管治和監察等問題，不少大廈的大維修花了一大筆金錢，亦不見得令建築物妥為修葺。

有意見促請政府考慮成立舊樓管理局，統合各相關部門直接處理舊樓的維修安全問題，包括強制出現安全問題的樓宇成立法團，或在過渡期直接委任管理公司或獨立非業主法團「接管」處理大廈的修葺工作。

退休保障

本地家庭通常會努力積儲部分收入，以應付失業或患病等的不時之需。此外，儲蓄亦可用作長遠的家庭計劃。雖然退休人士可以提取強積金資產及依靠家人供養，幫補部分生活費，但他們仍然需要有豐厚積蓄，才能維持自給自足的退休生活。

更值得令人關注的是，近年本港家庭債務急升，引起社會關注。由於預期美國利率即將上升，或會觸發本地物業市場出現調整，家庭債務的上升趨勢亦為整體經濟的潛在風險。

最新數字顯示，不少家庭在財務緊絀情況下借助家庭貸款渡過財政難關，這反映長遠的潛在風險，特別是在美國利率周期逆轉之後。

保安事務

香港警務署計劃為所有前線警務人員配備隨身攝錄機。警方指隨身攝錄機能讓警務人員以更客觀、問責及透明的方式搜證，從而提高執法成效。

然而，有人權組織關注到，由於警務人員可決定「是否錄影及何時開始錄影」，加上警務人員本身的行動不被攝錄，隨身攝錄機未必能夠全面及公平地呈現事件發生的經過。此外，隨身攝錄機可能收集多於執法所需的資料及數據，引發私隱保護問題的關注。

大體而言，根據海外經驗，警方以隨身攝錄機監察，拍攝所得的片段可減低證供上的爭議，加快解決投訴和訴訟；同時當公眾知道正被拍攝，他們通常會平靜下來，因而可減少警民衝突。

但值得關注的是，在警民互動的過程中，若警務人員有較大酌情權決定何時開啟及關掉攝錄機，或會增加警方使用武力的機會。

社會企業

社會企業是一盤生意，宗旨是要達致特定的社會目的，例如提供社會所需的服務或產品、為弱勢社群創造就業和培訓機會、保護環境，或運用本身賺取的利潤資助轄下的其他社會服務。社會企業所得的利潤主要用於再投資本身業務，以達到既定的社會目的營商宗旨，而並非為了賺取最大利潤。

香港社會企業的發展因缺乏營商經驗受到限制，政府現時為成立社會企業而設的資助計劃，僅供非牟利機構申請，而這等機構或缺乏商業經營的專業知識，其社會企業未必能夠在公開市場的競爭中生存。事實上，超過五分一的香港社會企業在耗盡種子基金後，被迫結業。

英國社會企業取得成功，可歸因於當地政府設立各種各類非政府中介機構，提供財務支援和專業意見，並協助這些社會企業吸引私人投資。

社會福利及服務

智障人士的家長關注其子女的福祉和經濟保障，尤其是當他們離世後，其子女的生活能否得到保障。在本港現時為精神上無行為能力人士而設的監護制度，成年人委任監護人管理現金資產最高限額為每月 15,000 港元。

部分家長認為獲委任的監護人的理財權力受到太大限制，每月發放的款額以 15,000 港元為上限未必足夠應付其智障子女的特殊需要。香港智障兒童的家長可考慮設立私人信託，在管理的資產取得較大的彈性，但設立私人信託涉及高昂的行政費用。

參考新南威爾斯州制度，政府為公眾人士提供全面的受託人服務，有特殊需要兒童的家長可選擇在遺囑中設立子女信託，家長亦可選擇在遺囑或信託契據中設立特殊殘障信託，受益人及合資格的家長和供款人可獲得社會保障經濟審查寬免。

體育發展

香港體育發展遠遜於其他鄰近地區，有指康文署及各體育總會應負上一定的責任，論據如下：

一、康文署缺乏明確客觀的準則以釐定體育總會獲撥的資助金額，資助金一直以機械化方式分配，少有參照體育總會的表現；

二、康文署沒有為員工提供適切培訓，導致該署職員未能有效監察各體育總會的表現，包括體育總會逾期提交報告及財務報表等；

三、雖然體育總會有僱用全職或兼職人員負責日常行政運作，但領導層仍然為非專職的義務人士。在這種管治模式下，即使政府增撥資源，亦難以改善體育總會的管治情況。

行政長官及國家元首政府首長

行政長官涉及利益衝突等問題，在社會上引起廣泛關注。行政長官、主要官員及公務員均受普通法中有關賄賂及公職人員行為失當罪行所規管，但仍有以下三點漏洞：

1. 《防止賄賂條例》適用於主要官員及公務員，但這條例的第 3 條及第 8 條，並不適用於行政長官。

2. 《公務員守則》，均有明確規則，就主要官員及公務員防止及處理利益衝突的情況作出規管，但現時未有規管行政長官。

3. 《政治委任制度官員守則》及一系列規管公務員的規條亦就主要官員及公務員接受利益時應注意的事項發出指引，但現時未有規管行政長官。

旅遊

「創意旅遊」屬新一代的旅遊模式，將旅遊由靜態的文化消費活動轉化為讓遊客參與旅遊目的地的文化創意活動，令其從中獲得互動體驗。南韓政府致力促進旅遊業與創意產業的融合，包括「粉絲見面會」、「韓食財團」、醫療旅遊、將藝術融入當地鄉村和傳統市場等。

在推廣創意旅遊方面同樣成功的其他例子包括巴西、新西蘭及泰國。巴西為遊客提供的體驗不只是觀看森巴舞表演，而是讓他們學跳森巴舞；新西蘭則提供由當地導師開辦的各種與土著文化相關的體驗工作坊；而泰國在鄉郊地區為遊客提供農舍生活體驗之旅，以及製作泰菜烹飪材料的工作坊。

在香港，政府有意發展文化創意旅遊，包括設於灣仔的「香港動漫海濱樂園」，和活化後的中區警署建築群的「大館」等。然而，部分議員認為發展文化創意旅遊的措施瑣碎，對提高整體旅遊業的競爭力幫助有限。

貿易

在香港，政府資助中小企在電子商貿平台進行出口推廣活動。然而，電子商貿在中小企的普及程度偏低，原因包括它們缺乏應用數碼方案的能力和知識。

在英國及德國等海外地方，當地政府採取更具策略性的方式推動電子商貿的應用，包括：

一、加強中小企業對數碼科技的認知和能力，令其更有信心轉用網絡銷售；

二、推出便利措施，優惠包括繳付較低佣金及享用「網站免費試用期」，與全球大型網上商店美國的亞馬遜（Amazon）及中國的天貓建立夥伴合作關係，鼓勵跨境電子商貿及拓展海外市場等。

交通運輸

最近數十年，全球已有多個地方實施一地兩檢的邊境管制安排。

一、深圳灣口岸：香港在深圳灣口岸的港方口岸區內享有全面的刑事及民事司法管轄權。雖然兩個司法管轄區的邊境管制人員身處同一樓宇，但他們在兩個毗鄰的區域各自執行與清關，管轄權沒有任何重疊；

二、英法海底隧道（Channel Tunnel）：乘坐歐洲之星往來英、法兩地城市的乘客，在設有一地兩檢安排的鐵路站啟程，僅須辦理出入境管制手續一次，無須在目的地車站再接受檢查；

三、美國與加拿大：美國預檢人員有權在執行職務時檢查、搜查及拘留任何人但美國人員不可攜帶槍械，有需要時則可使用非致命或不足以引致嚴重身體傷害的武力。值得注意的是：美國人員在預檢區內並無拘捕權，任何拘捕行動均須由加拿大人員根據加拿大法律進行。

邊境管制實施的一地兩檢安排可以多種形式出現，最重要的是雙邊或多邊達成共識。

廢物管理

現時本港每日棄置大量廚餘，其中約七成源自家居，其餘則源自工商業。政府即將推行廢物徵費，希望能夠減低廚餘數量，但業界認為作用不大，因為食肆根本不會為節省小量成本而大費周章減廢。

政府曾計劃興建現代化的廚餘處理設施，減少廚餘棄置到堆填區，但運作期間排出的污水，海水受污染會影響當地的養魚業及他們的生計。政府現正亦與工商界合作推行「廚餘循環再造合作計劃」，協助業界盡量避免及減少製造廚餘和在源頭分類，但作用一般。

有學者認為應立法監管廚餘廢物不能直接運往堆填區，以便更有效地完善廢物源頭分類，亦有意見認為應先設立配套措施以協助回收。

目前政府積極考慮透過焚化方案，減低全港的都市固體廢物量。

CHAPTER 4 英文模擬試卷

Background

Manboychun is a small Asian city with 800 million population. The four pillar industries, comprising trading and logistics, financial services, professional and producer services, and tourism, have been a driving force of the Manboychun economy for years. While the four pillar industries have been a driving force of Manboychun's economic growth for years, the Government has promoted six industries where Manboychun enjoys clear advantages for diversifying the local industry structure, including cultural and creative industries, medical services, educational services, environmental industries, innovation and technology, testing and certification services.

In the past few years, there is increasing interest in the concept of 'smart city' among governments, business and the general public around the world. Not only does it present opportunities to improve cities and quality of living, smart city development is expected to foster technology development and business growth, both local and international.

Development of Smart City in Manboychun

Manboychun is well-positioned to pursue smart city development with its advanced information and communications technology infrastructure, and is an early adoptor of Internet of Things technologies. Government departments as well as public and private organisations have been adopting sensors and related technologies in various fields to achieve their respective policy objectives. For instance, the Transport Bureau deploys sensors at busy roads for collection of real-time transport data, the Drainage Services Bureau uses intelligent ultrasonic sensors to detect water levels in manholes of different types of drains for prioritising maintenance and cleaning works to minimise flooding risks, and the Civil Engineering and Development Bureau uses sensors in monitoring slope conditions for landslide prevention. In addition, the Customs and Excise Bureau has deployed the E-Lock System as an efficient, secure and traceable customs clearance process which allows tracking through the Global Positioning System and reduce customs examinations across the boundary and clearance time significantly.

Definiton of a 'Smart City'

There is no universally adopted definition of "Smart City". However, based on the definitions adopted by different places and institutions, two common features of a "Smart City" are observed as follows: being a city that leverages on the information and communication technology infrastructure and uses innovative solutions to address issues in one or more aspects of the city including governance, economy, mobility, environment, living and people; and aiming at improving the quality of life of the citizens and enhancing the sustainable growth and competitiveness of the city through the "smart" initiatives.

Many major cities around the world such as Singapore and New York have introduced smart city projects with different objectives and priorities (Source A). Some cities, like Seoul, aim to make a broad range of city functions smart (Source B).

Consensus among stakeholders

The majority of senators had passed the following bills and motions:

(a) Public-private partnership: smart city calls for close collaboration among the public sector, private sector, academia and citizens over the whole cycle of implementation: from identification of city challenges, formulation of policy and strategy, research and development, conceptualisation of potential projects, feasibility assessment, proof of concept through pilot projects to city-wide implementation.

(b) Smart Region Living Lab: new technology solutions can be tested in a special environment before their wider adoption in the city, e.g. autonomous vehicle would be tested in the restrinformation and communications technologyed areas of the Manboychun International Airport and other suitable locations

To achieve the above tasks, the digital infrastructure in Manboychun will be upgraded and updated in the following means:

(a) enhancing city connectivity through expanding the coverage of "Wi-Fi.Manman" by doubling the number of hotspots, doubling the speed of Wi-Fi connection at government venues and strengthening security (Source C);

(b) The administration proposes setting up a one-stop online eID system, as a key digital infrastructure for smart city, to enable residents to use electronic services and conduct online transactions in a more convenient and secure manner. The proposed eID could be regarded as a common key for digital identity authentication for members of the public to login and access various government and commercial electronic services in a simple and secure manner (Source D).

Objective

In the 2020 Policy Address, the Prime Minister reaffirmed the Administration's commitment "to developing Manboychun into a smart city by using innovation and technology to enhance city management and improve people's livelihood". The Administration's objective of smart city development is to inspire continuous city innovation and sustainable economic development. In total, there are two pilot schemes suggested for the consideration of the Government.

Pilot scheme 1: Smart public transport interchanges

Major bus stops will be transformed into smart public transport interchanges with the provision of 3rd party hardware, access of free Wi-Fi and multi-purpose touch screen offering a range of information (e.g. news, weather conditions, estimated time of arrival of the buses to the bus stops, the passenger' s estimated time of arrival to the destination based on traffic). There are also integration of sensors into poles to demonstrate multi-functionalities (e.g. traffic detection, air quality) and information and communications technology enablement for data collection. Additionally, there are other potential functions to consider include 999 Emergency Call button which links to a video camera for capturing real-time situation when the button is pressed and broadcasting of emergency messages.

Pilot scheme 2: Cultural and heritage tourism

Given the importance of the tourism industry in Manboychun it is worthwhile to explore wider use of information and communications technology to further enhance tourists' experiences by digital elements like interactive videos or photos, electronic coupons or discount offers at selected tourist spots, digital walking routes guide and on-line information on accessibility and other matters. AR and VR developers and user experience specialists may also partner with the Government to provide the suggested interactive content. The pursuit of technology-oriented projects will open opportunities for partnership in innovation and local app developer, local device and solution providers. Tailored coupons and increased marking will improve business to retailers, restaurants and other business operators in the tourist spots.

Task

Different cities have different strategies in pursuing smart city development, having regard to their respective policy, local circumstances and resources. It is imperative for the administration to first formulate a Smart City Development Initial Pilot Scheme for Manboychun.

You are the officer responsible for smart city development in Manboychun. Please draft a discussion paper for the Prime Minister's Urban Development bureau, setting out the pros and cons of the pilot scheme (i) and (ii), and all other relevant considerations. Which options do you think government should go with? Explain your answer with justifications. Apart from the options listed, you may suggest new policy initiatives.

Source A: objectives of Smart City development of major cities

Major cities	Development
Barcelona	integrate town planning and extensive use of information and green technology
Copenhagen	build a green city and seeks to become a technology solutions leader
New York	adopt innovative service models in open data sharing
Singapore	build a well-connected city by developing and employing information and communications industries
Vienna	achieve growth coupled with reduced resource consumption through innovations

Source B: details of smart city development by Seoul Metropolitan Government

Areas	Goal	Technical details
Smart government initiatives	enhance citizens' convenient access to efficient and innovative public services and promote transparency of its administration	(i) upgrading its digital public services through the use of big data, delivering public services on mobile devices (ii) sharing of data with citizens as much as possible based on the principle of open data
Smart living	promote energy conservation, citizens' safety and efficient transport services	1,000 households were installed with smart meters and provided with real time reports of electricity, water and gas consumption in terms of monetary units. The initiative might contribute a 10% reduction in energy consumption among these households.
Smart economy	supported businesses that develop new technologies of information security essential for smart phones, CCTV and cloud computing	(i) increase the investment in strengthening security for mobile public administration and other applications using smart technologies (ii) promote application businesses by nurturing application experts and supporting entrepreneurs

Source C:

To enhance city connectivity, the Government has pledged to progressively expand the coverage of "Wi-Fi.Manman" by doubling its number of hotspots to 50 000 within three years, covering venues including public rental housing estates, public hospitals, markets, parks, sitting-out areas, promenades, tourist spots, public transport interchanges and land boundary control points, etc. The speed of Wi-Fi connection at government venues will be doubled progressively to 3-4 Mbps and its security will also be further enhanced.

In addition, the administration will offer free Wi-Fi services at all youth service centres and study rooms run by the Government and non-profit-making organisations. Furthermore, the administration will collaborate with public and private organisations to expand "Wi-Fi.Manman" coverage to venues of high public patronage such as busy streets, bus stops and shopping malls.

Source D:

The eID would be made available for free for all Manboychun residents to apply and use on voluntary basis. For e-government services, eID can be connected to different services. For example, it can be used to submit online applications for services like licence renewal, booking of venues, making appointments, etc., or document signing; authorise retrieval of information stored in electronic service system to pre-fill online forms or update address; and, by way of a unified identity authentication, facilitate the development of cross-departmental or institutional electronic services and streamline business processes.

Currently, government and commercial e-service providers are using different authentication systems to verify users' identity, such as username and password, secure token, SMS, etc. With eID, residents can access various electronic services by way of a single login anytime and anywhere, so that they can enjoy consistent user experience, and avoid the inconvenience of managing different groups of usernames and passwords or carrying multiple secure tokens, thus bringing convenience to their daily living.

若考生是last minute才看這書的，這裏要提提您，不論您是選擇哪一個方案，甚或至建議一個新方案，都要詳述「考慮準則」，否則必定「肥佬」，這些準則包括：

1. Can be implemented in a relatively short timeframe;
2. Are highly relevant to the community;
3. Bring tangible benefits to the community;
4. Have the potential for territory-wide implementation;
5. Are likely to be sustainable in the long term.

CHAPTER 5 中文模擬試卷

近年，不時有人批評行政機關沒有按用人唯才的原則委任諮詢及法定組織的成員。

有人建議當局改革諮詢及法定組織成員的委任機制，規定一些主要職位的委任須獲專責委員會經聆訊後通過。試分析上述建議的三個利弊。

若你同意上述建議，請闡述兩點執行細節。

若你不同意上述建議，請建議兩個方法改善現況。

為保障讀者權益，請把收據電郵至 leesirjre@gmail.com ，索取評卷準則（Suggested marking guidelines）。若考生是最後一分鐘才看這書，筆者要提提您。不論您的立場如何，都必須加入大量「就你所知」的關於香港情況的概念和知識，否則必肥！

參考答題(節錄)

1. 定義

透過政府的諮詢及法定組織，社會各界人士和有關團體可在政府制訂政策和籌辦公共服務給予意見，及參與各種公共事務。

2. 三個利弊(舉例:弊處)

專責委員會經的聆訊程序未必能顧及有關組織的需要外，因這可能阻礙政府羅致不同背景和經驗的人士，例如專業人士、學者、工商界人士、地區及有關界別的代表等。

3. 表達立場(舉例:反對)

我認為目前諮詢及法定組織非官方成員的委任制度基本上運作暢順及有效。我建議政府各政策局及部門繼續發掘更多來自不同界別、有才幹人士參與這些組織，讓他們可以發揮所長，協助政府掌握社情民意，有效施政。

JRE
急症室

後記

此書愈來愈好賣，口碑甚佳，聽聞最大功勞不是出版社，而是考題官！

據說政務官離職潮下，政府不斷在JRE考試中錄用大量考生為AO。唯此舉仍未能解決AO人手不足問題，近年政府更要舉行多場政務官內部招聘程序，從其他職系（例如EO、CO等）「搶人」來做AO。

然而，內部招聘並非政府首選做法，特區仍然希望是從JRE考試中取錄AO的。有見及此，江湖傳聞聘用組的同事，都紛紛向學弟學妹、身邊朋友推介一定要報讀JRE，一定要溫書，做足準備功夫投考政府。

至於溫書方法是什麼？筆者不知道。唯筆者發現在截稿前，各大書局只有這本JRE　Textbook。又，近年JRE的題目、題型、考核模式等跟本書的教學內容非常相近，尤其是中文部分的五十多個熱門議題哩！

JRE 考生編號

你可以把自己的年齡、身份證號碼、信用卡密碼告知他人，但絕不可把 JRE 的考生編號洩露出去。

官方規定，若發現兩份試卷的考生編號相同，無須審判，兩者即時 DQ。為甚麼局方不作調查？公務員的惰性，你要學得懂才能當官喔！

大家都能想像：每年都有人報考 JRE ，半小時後便早走，為的便是寫上仇家的考生編號。報 JRE 不收報名費，只需安坐半小時，便破壞了人家的一生前途。不是一年的前途，而是一生的前途，因為有些人只要 JRE 衰過，便不會再考。

雖然每次 JRE 收卷局方均指示公務員要核對考生編號，但效果如何……相信你亦能猜到。

所以 JRE 不合格，可能錯不在你。

請槍代考？

「請槍代考」是非常普遍的事。在香港中學文憑試中，考場設閉路電視，各監考員設電子儀器檢查考生文件，意圖盜絕代考事宜。

至於香港 JRE 考試，監考員只會用手拿起身份證，目測確認該考生身份。香港假身份證難做嗎？最難的部分是那張晶片，基本上任何普通儀器一 Check，便知是假身份證。

但問題是，香港 JRE 考試，監考員沒有相關儀器檢測晶片，只能通過手感和觀察身份證照片。

每年有多少代考呢？我沒有數字，但每年 JRE 極高分，文法全對，但在面試時英文文法錯漏百出，講嘢低能的例子倒也聽過不少。

Hygenie break

考試時間為三小時，中間沒有小息，考生去洗手間是平常之事。監考員會陪同考生到洗手間門口。不需目測，只需聆聽便可得知是否有手機或紙張聲音。一個考生要找資料，總需五分鐘吧，一般考生在三分鐘內出來的，怎會作弊呢（常人道）。

若大家有監考 JRE 經驗，或多次應考 JRE ，都會發現一部分考生會完成英文卷後去洗手間，而每年坊間都有同一個問題，為甚麼先做英文卷？

一般的解釋是，英文卷是 Data-based questions，只需抄寫便可；相反中文卷只有問題數句，考生需多加思考論點和論據，於是考生先完成英文卷，再看畢中文卷題目，在往返洗手間圖中，便可不斷諗 Point，不浪費考試的一分一秒。

另一個解釋是，有考生進場閱卷後，半小時後離場，之後在網上搜集相關資料，再放到各試場的洗手間（位置不便透露），考生便可一至兩分鐘 hygiene break 閱畢整條題目的論點，之後返回試場 elaborate。

你相信哪個版本？其實不必積極留意是否有一大堆考生完成英文卷後去洗手間，而是看清楚這本書貼的 50 多個中文熱門議題更實際。

參考資料：

1. 香港特區政府一站通

2. 香港立法會會議紀錄

3. 香港立法會秘書處研究資料

4. 香港考試及評核局出版刊物

5. 戰勝 GRPT 全民英檢中級的 16 課

書名　投考公務員 AO/EO JRE 全攻略（增修第五版）
ISBN　978-988-76628-8-4
定價　HK$148
出版日期　2023年6月
作者　李 Sir
責任編輯　文化會社投考公務員系列編輯部
版面設計　梁文俊
出版　文化會社有限公司
電郵　editor@culturecross.com
網址　www.culturecross.com
發行　聯合新零售（香港）有限公司
地址：香港鰂魚涌英皇道1065號東達中心1304-06室
電話：（852）2963 5300
傳真：（852）2565 0919